Lecture Notes in Bioinformatics 9702

Subseries of Lecture Notes in Computer Science

More information about this series at http://www.springer.com/series/5381

María Botón-Fernández · Carlos Martín-Vide
Sergio Santander-Jiménez
Miguel A. Vega-Rodríguez (Eds.)

Algorithms for Computational Biology

Third International Conference, AlCoB 2016
Trujillo, Spain, June 21–22, 2016
Proceedings

 Springer

Editors
María Botón-Fernández
CETA-Ciemat
Trujillo
Spain

Carlos Martín-Vide
Rovira i Virgili University
Tarragona
Spain

Sergio Santander-Jiménez
University of Extremadura
Cáceres
Spain

Miguel A. Vega-Rodríguez
University of Extremadura
Cáceres
Spain

ISSN 0302-9743   ISSN 1611-3349   (electronic)
Lecture Notes in Bioinformatics
ISBN 978-3-319-38826-7   ISBN 978-3-319-38827-4   (eBook)
DOI 10.1007/978-3-319-38827-4

Library of Congress Control Number: 2016938660

LNCS Sublibrary: SL8 – Bioinformatics

Printed on acid-free paper

This Springer imprint is published by Springer Nature
The registered company is Springer International Publishing AG Switzerland

Preface

These proceedings contain the papers that were presented at the Third International Conference on Algorithms for Computational Biology (AlCoB 2016), held in Trujillo, Spain, during June 21–22, 2016.

The scope of AlCoB includes topics of either theoretical or applied interest, namely:

- Exact sequence analysis
- Approximate sequence analysis
- Pairwise sequence alignment
- Multiple sequence alignment
- Sequence assembly
- Genome rearrangement
- Regulatory motif finding
- Phylogeny reconstruction
- Phylogeny comparison
- Structure prediction
- Compressive genomics
- Proteomics: molecular pathways, interaction networks
- Transcriptomics: splicing variants, isoform inference and quantification, differential analysis
- Next-generation sequencing: population genomics, metagenomics, metatranscriptomics
- Microbiome analysis
- Systems biology

AlCoB 2016 received 23 submissions. Most papers were reviewed by three Program Committee members. There were also several external reviewers consulted. After a thorough and lively discussion phase, the committee decided to accept 13 papers (which represents an acceptance rate of about 56 %). The conference program included three invited talks and some presentations of work in progress as well.

The excellent facilities provided by the EasyChair conference management system allowed us to deal with the submissions successfully and handle the preparation of these proceedings in time.

We would like to thank all invited speakers and authors for their contributions, the Program Committee and the external reviewers for their cooperation, and Springer for its very professional publishing work.

March 2016
María Botón-Fernández
Carlos Martín-Vide
Sergio Santander-Jiménez
Miguel A. Vega-Rodríguez

Organization

AlCoB 2016 was organized by the Computer Architecture and Logic Design Group (ARCO) of the University of Extremadura, Cáceres, the Extremadura Centre for Advanced Technologies (CETA-CIEMAT), Trujillo, and the Research Group on Mathematical Linguistics (GRLMC) of Rovira i Virgili University, Tarragona.

Program Committee

Can Alkan	Bilkent University, Ankara, Turkey
Timothy L. Bailey	University of Queensland, Brisbane, Australia
Vladimir Bajic	King Abdullah University of Science and Technology, Thuwal, Saudi Arabia
Geoff Barton	University of Dundee, UK
Inanc Birol	University of British Columbia, Vancouver, Canada
Jacek Blazewicz	Poznan University of Technology, Poland
Alan P. Boyle	University of Michigan, Ann Arbor, USA
Vladimir Brusic	Nazarbayev University, Astana, Kazakhstan
Liming Cai	University of Georgia, Athens, USA
Rita Casadio	University of Bologna, Italy
Ken Chen	University of Texas MD Anderson Cancer Center, Houston, USA
Jason Ernst	University of California, Los Angeles, USA
Laurent Gautier	Novartis Institutes for BioMedical Research, Cambridge, USA
Manolo Gouy	Claude Bernard University of Lyon 1, France
Michael Gribskov	Purdue University, West Lafayette, USA
Iman Hajirasouliha	Stanford University, USA
John Hancock	Genome Analysis Centre, Norwich, UK
Artemis Hatzigeorgiou	University of Thessaly, Volos, Greece
Fereydoun Hormozdiari	University of California, Davis, USA
Kazutaka Katoh	Osaka University, Japan
Evangelos Kranakis	Carleton University, Ottawa, Canada
Lukasz Kurgan	University of Alberta, Edmonton, Canada
Bill Majoros	Duke University, Durham, USA
Lennart Martens	Ghent University, Belgium
Maria-Jesus Martin	European Bioinformatics Institute, Hinxton, UK
Carlos Martín-Vide (Chair)	Rovira i Virgili University, Tarragona, Spain
Folker Meyer	Argonne National Laboratory, USA
Kenta Nakai	University of Tokyo, Japan
Matteo Pellegrini	University of California, Los Angeles, USA
Mihaela Pertea	Johns Hopkins University, Baltimore, USA
Ben Raphael	Brown University, Providence, USA

Paolo Ribeca	Pirbright Institute, Woking, UK
Denis Shields	University College Dublin, Ireland
Fredj Tekaia	Pasteur Institute, Paris, France
Alessandro Verri	University of Genoa, Italy
Fuli Yu	Baylor College of Medicine, Houston, USA
Daniel Zerbino	European Bioinformatics Institute, Hinxton, UK
Kaizhong Zhang	University of Western Ontario, London, Canada
Weixiong Zhang	Washington University in St. Louis, USA
Zhongming Zhao	Vanderbilt University, Nashville, USA
Yaoqi Zhou	Griffith University, Brisbane, Australia

External Reviewers

Josep Basha
Zechen Chong
Liang Ding
Xian Fan
Luca Ferretti
Dimitrios Kleftogiannis
Yi Liu
Mohammad Mohebbi
Ibrahim Numanagic
Sung-Joon Park
Victoria Popic
David Robinson
Hsin-Ta Wu
Yan Yan

Organizing Committee

María Botón-Fernández	Trujillo (Co-chair)
Carlos Martín-Vide	Tarragona (Co-chair)
Miguel A. Vega-Rodríguez	Cáceres (Co-chair)
Florentina Lilica Voicu	Tarragona

Local Committee

Leslye Alarcón	Trujillo
María Botón-Fernández	Trujillo (Co-chair)
José M. Granado-Criado	Cáceres
Sergio Santander-Jiménez	Cáceres
Miguel A. Vega-Rodríguez	Cáceres (Co-chair)

Contents

Sequence Analysis and Rearrangement

Invited Talk

The Trees in the Peaks

David Sankoff[1]([⊠]), Chunfang Zheng[1], Eric Lyons[2], and Haibao Tang[3]

[1] Department of Mathematics and Statistics, University of Ottawa,
585 King Edward Avenue, Ottawa K1N 6N5, Canada
{sankoff,czhen033}@uottawa.ca
[2] School of Plant Science, Bio5 Institute, University of Arizona,
Tucson, AZ 85721, USA
ericlyons@email.arizona.edu
[3] Center for Genomics and Biotechnology,
Fujian Agriculture and Forestry University, Fuzhou 350002, China
tanghaibao@gmail.com
http://albuquerque.bioinformatics.uottawa.ca

Abstract. We suggest a gene-tree/species-tree approach to speciation and whole genome duplication (WGD) to resolve the occurrence of these events in phylogenetic analysis. We propose a more principled way of estimating the parameters of gene divergence and fractionation than the standard mixture of normals analysis. We formulate an algorithm for resolving data on local peaks in the distributions of duplicate gene similarities for a number of related genomes. Illustrating with a comprehensive analysis of WGD-origin duplicate gene data from six members of the family Brassicaceae, we discuss the effects of variable evolutionary rates and data degradation due to fractionation. We introduce the notion of peak tree, as a template for all gene trees evolving by speciation, WGD and fractionation.

Keywords: Gene tree · Species tree · Whole genome duplication · Algorithms · Mixture of distributions · Brassicaceae

1 Introduction

The investigation of gene trees and species trees furnishes a genomic perspective on evolution insofar as it requires a complete inventory of the paralogs of the orthologously related genes in the species under study. This line of study also requires a different king of algorithm than those familiar from traditional single-gene based phylogenetics, or even the so-called "phylogenomics" based on large numbers of concatenated genes using what is basically traditional methodology. However, gene trees and species trees are each based on a tiny portion of the genome. In the context of whole genome duplication (WGD) in flowering plants, we can take the gene-tree/species tree approach to a more comprehensive kind of genomic data than the usual one-gene-at-a-time focus.

Specifically, we will study the set of $\binom{N}{2} + N$ gene similarity distributions within and across N species where WGD has affected one or more of these

© Springer International Publishing Switzerland 2016
M. Botón-Fernández et al. (Eds.): AlCoB 2016, LNBI 9702, pp. 3–14, 2016.
DOI: 10.1007/978-3-319-38827-4_1

species. This typically involves many thousands of genes. This paper raises more technical problems than it solves, but its goal is to show how concepts from gene-tree theory enable us to better understand genomic history.

We first sketch out a model of gene similarity distribution under random sequence divergence, speciation and fractionation, leading to a principled treatment of the statistical inference of divergence and fractionation rates and to speciation and WGD times.

Still lacking an implementation of this methodology, we can nonetheless proceed with our gene-tree approach by simply identifying local modes or "peaks" in all the similarity distributions, and translating these into phylogenetically related paralogous and orthologous entities. We present a rapid algorithm to resolve these in the case of ideal instances where no data are missing and all data are mutually compatible.

Finally, we illustrate our approach with six species spanning three genera of the family Brassicaceae.

2   Distributions of Gene Similarity

2.1   Background

We will discuss the distribution of similarities between homologous genes, according to a simple model that takes into account only

- gene mutation by random substitution of nucleotides independently at each position, and
- random duplicate gene loss after whole genome duplication (WGD).

Moreover, to simplify we treat all genes as having length l, i.e., l positions each containing one nucleotide.

After speciation, the genes in the two new species diverge independently according to a rate parameter λ. The simplest model for this divergence is based on binomial trials for change of nucleotide at each of the l positions of the gene. A success in the binomial trial at a position is the event that the nucleotide is the same at time t in both species. The similarity of the pair of orthologous genes at time t is binomially distributed $B[l, p(\lambda, t)]$, where $p(\lambda, t) = e^{-2\lambda t} + \frac{1}{4}(1 - e^{-2\lambda t})$. As time elapses, $p \to \frac{1}{4}$, so that the similarity between genes becomes indistinguishable from "noise", since $p = \frac{1}{4}$ is characteristic of pairs of random sequences.

Since we treat all n genes as having l nucleotides, the predicted frequency distribution of successes is given by $\Phi[n, \lambda, t] = nB[l, p]$. Inference under this model is simple. If the empirical frequency distribution of similarities (number of successes across l trials) is nF, where the mean of F is m, then $\hat{p} = \frac{m}{l}$ and $\hat{\lambda} t = -\frac{1}{2} \log \frac{4\frac{m}{l} - 1}{3}$. If t is known, this gives us an estimate of the mutation rate constant λ, while if λ is known, this gives us an estimate of the divergence time t.

When a genome undergoes whole genome duplication (WGD), each gene is duplicated, creating one pair of "paralogous" genes. Over time, the frequency

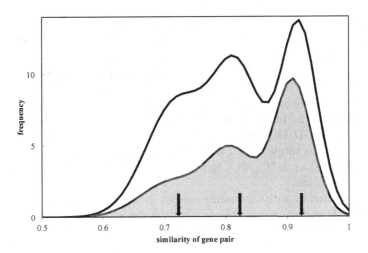

Fig. 1. Idealized gene similarity distribution between two species represented by three events, two WGD and a speciation. The arrows indicated $p = exp-\lambda t$ for the individual events. The number of gene pairs throughout is 4000. The upper curve represents the situation where the fractionation rate ρ is zero; the broadening of the part of the distribution reflecting the earliest event is due to increasing variance with greater age. The lower curve bounding the shaded area adds the effect of non-zero ρ so that the earliest event is increasingly hidden by subsequent events. The x-axis in this type of diagram is often shown in a log scale, scaling linearly with time, so that very early events appear much farther to the left.

distribution of the similarities between paralogous genes becomes $e^{-\rho t}\Phi[n, \lambda, t]$, where the rate parameter ρ accounts for process of losing (deleting or inactivating) one of each duplicate gene pair. Note that paralogous genes begin diverging at a duplication event, while orthologous genes begin diverging at a speciation event. Inference of ρ and λ, or of t, from frequency data is once again straightforward.

Speciation and WGD events can combine in any number of ways in the history of an evolutionary domain. For example, after speciation, one of the two sister genomes may undergoes WGD at time t_1, where $0 \leq t_1 \leq t$. Here, the similarities between the $2n$ pairs of homologs in the two species at time t, will be distributed as $2e^{-\rho[t-t_1]}\Phi(\lambda, t)$. Again there is no difficulty in inferring the parameters.

If however after WGD at time 0, a genome undergoes speciation at time t_1, where $0 \leq t_1 \leq t$, then the similarities between the $4n = 2 \times 2n$ pairs of homologs in the two species at time t, will be distributed as $2e^{-\rho t}[\Phi(\lambda, t) + \Phi(\lambda, t - t_1)]$. (This distribution is bi-modal when the parameters λ and ρ are suitably small with respect to t_1 and $t - t_1$.) In this model there is no closed form for the maximum likelihood estimators of the parameters or times. It is the usual practice to resort to numerical procedures embodied in software such as EMMIX [10] for resolving mixtures of normal distributions.

Similarly, if after whole genome triplication (WGT), a genome undergoes speciation at time t_1, where $0 \leq t_1 \leq t$, the similarities between the $9n = 3 \times 3n$ pairs of homologs in the two species at time t, will be distributed as $3e^{-\rho t}[2\Phi(\lambda, t) + \Phi(\lambda, t - t_1)]$. The same kind of logic applies to a speciation after a higher degree polyploidization (whole genome quadruplication, etc.) These models all have the same inferential complications as the previous one. They can all also result in bimodal distributions, In the case a genome undergoes two successive WGD at time 0 and t_1, then a speciation at time t_2, so that $0 \leq t_1 \leq t_2 \leq t$, the $16n = 4 \times 4n$ pairs of homologs in the two species at time t, will be distributed as $4e^{-\rho t}[2\Phi(\lambda, t) + \Phi(\lambda, t - t_1) + \Phi(\lambda, t - t_2)]$. This can be a trimodal distribution.

It is important to note that many species in a phylogeny may be related by the same WGD and speciation events, and the times estimated for these events should be constrained to be equal. Other events should be constrained to occur in an order compatible with the phylogeny. Such constraints are not available in standard statistical mixtures of distributions software.

Our theoretical considerations pertain to the simple model assumed at the beginning of this section. In practice, various other processes affect the distribution of similarities so that the number of gene homologs between and within genomes may be severely reduced from those expected from the model. Within a group of related organisms, however, the parameter λ tends to have a constant value, although there are particular cases where it may be substantially lower [11] or higher [1]. The hypothesis of constant ρ has been investigated in [12].

The broadening of effects of event age and of fractionation on similarity distributions as time elapses are illustrated in Fig. 1. Eventually, all events become indistinguishable from noise caused by random gene resemblances, widespread domain sharing, tandem and near-tandem duplications, gene-order rearrangements, gene conversion and other processes.

It is important to note that methods like EMMIX, powerful and flexible as they may be, are not tailored to the problem of detecting speciation and WGD in a *set* of related similarity distributions. For any mixture of normals, EMMIX will identify these components as long as there is enough data. But not every mixture of normals credibly reflects some sequence of genomic events. More important, among the $\binom{N}{2} + N$ gene similarity distributions within and across N species, there are many constraints that are not handled by software packages, such as requiring $\hat{t}$ to be the same for an event in all the distributions that are affected by it.

Despite these problems with speciation- or WGD-event detection, in this paper, we will assume constant λ, and we will assume that we can infer p, and hence the age of an event, simply by identifying the mode, or "peak" of the similarity distribution, without recourse to other estimation procedures. This unfortunately foregoes any attempt at present to pick out events visible as "shoulders" of other events on the similarity distribution, but it will allow us to validate the notion that $\hat{t}$ should be the same for an event for all the distributions in which it plays a role.

2.2 From Peaks to Species Trees and Duplication Gene Trees

There are two observations underlying our method for reconstructing the species tree and "peaks tree" from a perfect set of inter- and intra-genome comparisons:

- each intra-genome distribution of similarities only has peaks due to all the WGD in its direct lineage, and
- each inter-genome distribution may contain many peaks due to WGDs, but only one peak due to speciation, i.e., at the date of the most recent common ancestor of the two species.

We use these principles one after the other to produce our results. The pseudocode below assumes a perfect set of inter- and intra-genome comparisons, namely that all events affecting a between-genome or within-genome comparison are detected by the kind of inferential statistics mentioned in Sect. 2, and these comparisons are found in all the genomes affected by the event, and only these, according to the above principles.

For an event i at time t_i, we write $(time, genome_1, genome_2)$. Each event time is associated with two genomes, which may be distinct or identical.

Algorithm 1. Construct the tree

Input: A set of genomes $G = \{g_1, g_2, \ldots, g_m\}$,
         A set of event times $E = \{t_1, t_2, \ldots, t_n\}$,
Output: A speciation tree in Newick format with duplication nodes

1 **for** $i \leftarrow 1$ **to** m **do**
2 ⌊ get all the duplicate time(s) Dt_i for genome i by **Algorithm 2**
3 **for** $i \leftarrow 1$ **to** $(m-1)$ **do**
4 ⌊ **for** $j \leftarrow i+1$ **to** m **do**
5 ⌊ ⌊ get the speciation time St for g_i and g_j by **Algorithm 3**
6 split G into G_{left} and G_{right} by **Algorithm 4**
7 **while** G_{left} *or* G_{right} *contains* ≥ 2 *genomes* **do**
8 ⌊ recursively apply **Algorithm 4** to G_{left} or G_{right}
9 **return** G *in the Newick format*

Algorithm 2. Get duplication time for a genome

Input: A genome g_i
         A set of event times $E = \{t_1, t_2, \ldots, t_n\}$,
Output: duplication event(s) Dt_i for g_i

1 $Dt_i \leftarrow \emptyset$
2 **for** $j \leftarrow 1$ **to** n **do**
3 ⌊ **if** $t_j : genome1 = g_i$ *and* $t_j : genome2 = g_i$ **then**
4 ⌊ ⌊ add $t_j : time$ to Dt_i
5 **return** Dt_i

Algorithm 3. Get speciation time for g_i and g_j

    Input: Two genomes g_i and g_j,
          Dt_i and Dt_j
          A set of event times $E = \{t_1, t_2, \ldots, t_n\}$,
    Output: Speciation time St for g_i and g_j

1 **for** $k \leftarrow 1$ **to** n **do**
2   **if** $t_k : genome1 = g_i$ and $t_k : genome2 = g_j$ **then**
3   **if** $t_k : t \notin Dt_i$ and $t_k : t \notin Dt_j$ **then**
4           $St : time \leftarrow t_k : t$
5           $St : genome1 \leftarrow g_i$
6           $St : genome2 \leftarrow g_j$
7 **return** St

Algorithm 4. Split a group of genomes into two groups by a SpeciationNode

---

    Input: A set of genomes ψ, can be G or subset of G
    a set of speciation times $\{St_1, St_2, \ldots, St_r\}$, for all pairwise genomes in ψ
    Output: A speciationNode and two subsets of ψ, ψ_{left} and ψ_{right}.
          $\psi_{left} \cup \psi_{right} = \psi$

1 $\psi_{left} \leftarrow \emptyset$
2 $\psi_{right} \leftarrow \emptyset$
3 $leftGenome = 0$
4 $rightGenome = 0$
5 $speciationNode = 0$
6 $duplicationNode = 0$
7 **for** $k \leftarrow 1$ **to** r **do**
8   **if** $St_r : time > speciationNode$ **then**
9        $speciationNode = St_r : time$
10       $leftGenome = St_r : genome1$
11       $rightGenome = St_r : genome2$
12 **for** *all the duplication times for each genome in* ψ **do**
13   **if** $\exists$ *a duplication time dt* $<speciationNode$ *AND all the genomes in* ψ
      have this duplication time **then**
14   duplicationNode = dt **for** *each genomes in* ψ **do**
15   remove dt from Dt of this genome
16 **for** $k \leftarrow 1$ **to** r **do**
17   **if** $St_r : time = speciationNode$ and
      $St_r : genome1(/genome2) = leftGenome$ **then**
18   Add $St_r : genome2(/genome1)$ to ψ_{right}
19   **if** $St_r : time = speciationNode$ and
      $St_r : genome1(/genome2) = rightGenome$ **then**
20   Add $St_r : genome2(/genome1)$ to ψ_{left}
21 **return** $duplicationNode, speciationNode, \psi_{left}, \psi_{right}$

---

3   The Brassicaceae

To illustrate our discussion, we draw on six published genomes in the Brassicaceae family, three in the genus *Brassica*: *B. rapa* (turnip, Chinese cabbage) [13], *B. oleracea* (cabbage, cauliflower) [7] and *Raphanus sativus* (radish) [6], two in the genus *Arabidopsis*: *A. lyrata* (rock cress) [3] and *A. thaliana* (thale cress, mouse-ear cress) [4] and one in the genus *Sisymbrium*: *S. irio* (London rocket) [2]. Figure 2 shows the phylogenetic relationship among the six species:

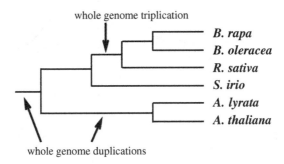

Fig. 2. Phylogenetic relationship of six species in the family Brassicaceae, showing lineages affected by WGD and WGT events.

We extracted genomic data from these species using the database in CoGe [8,9]. We then used the SynMap routine (with default parameters) on this platform to compare the gene orders of each of the $\binom{6}{2} = 15$ pairs of genomes. This procedure implicitly validates the identification of orthologs produced by speciation by detecting collinear arrays of several duplicate pairs in two species with approximately the same divergence: "syntenic blocks". Similarly, we did a self-comparison of five of the six genomes; the sixth one, the *Sisymbrium* genome, did not have enough closely spaced duplicate pairs for SynMap to produce paralogous syntenic blocks. The distributions of similarities calculated are shown in Fig. 3. The peaks found in each genome are tabulated in Table 1.

From Fig. 3 and Table 1, we note that the data are not quite "perfect"; the earliest duplication, detected at 79–80 % in the *Arabidopsis* self-comparisons, shows no peaks in the other self-comparisons – there is a shoulder or heavy tail in the appropriate place in the *Brassica* self-comparisons, but this is swamped by the later triplication. The triplication itself is visible in all three *Brassica* self-comparisons and in the comparison of *B. oleracea* and *B. rapa*, but not in the weaker signals involving *Raphanus*. Most of these missing data could be recovered using statistical means such as those discussed in Sect. 2 involving constraints instead of relying on identification of peaks.

More interesting is that the peaks at 90 % reflecting the *Sisymbrium* speciation, known to occur before the *Brassica* triplication, suggest that speciation is more recent, since the triplication peak is at 89 %. This apparent conflict is

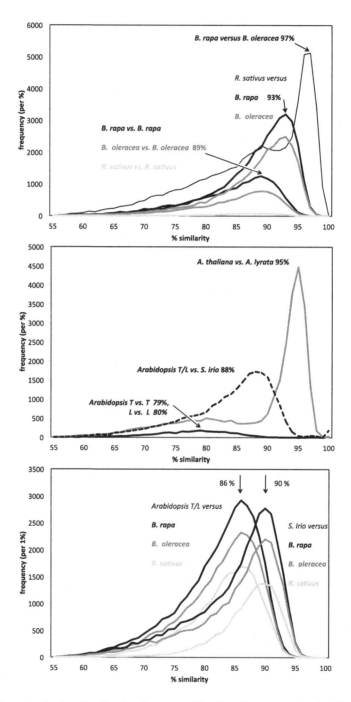

Fig. 3. Gene similarity distribution between 15 pairs of genomes in the Brassicaceae and 5 self comparisons. Local modes ("peaks") are indicated. Only one of each comparison is shown for *Arabidopsis*, the other is superimposed and indistinguishable.

Table 1. Peak similarity level, by genome. np: no peak, but one could be found by mixtures of distribution methods. - : no peak expected. Note peak 3 occurring before peak 4 due to slow evolutionary rate (λ) of *Sisymbrium*.

Peak number	Description	Genome					
		BR	BO	RS	SI	AL	AT
1	Alpha duplication [5]	np	np	np	np	80	80,79
2	Divergence of genus *Arabidopsis*	86	86	86,87	88	88-86	88-86
3	Whole genome triplication	89	89	87	-	-	-
4	Divergence of genus *Sisymbrium*	90	90	90	90	-	-
5	Divergence of genus *Raphanus*	93	93	93	-	-	-
6	Speciation of *Arabidopsis T & L*	-	-	-	-	95	95
7	Speciation of *B. rapa & B. oleracea*	97	97	-	-	-	-

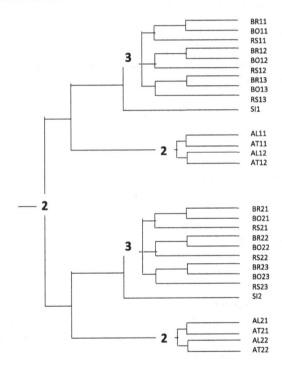

Fig. 4. Peak-tree for the Brassicaceae. Boldface numbers indicate WGD or triplication. All families of genes descended from the various genome WGD or WGT without any additional duplications must be formed from this tree with truncations of appropriate lineages.

clearly ascribable to a slower rate of evolution (lower λ), since the divergence of *Arabidopsis* from *Sisymbrium* also seems to occur more recently (88 %) than the divergence of *Arabidopsis* from the *Sisymbrium* sister genus *Brassica* (86 %). Note that the small differences between peak similarities are not insignificant, given the many thousands of gene pairs involved in these comparisons.

Were we to fill in the missing peaks, and correct the *Sisymbrium* times to account for slower evolution, the data set would be perfect and Algorithm 1 would convert it to a species tree with duplication times indicated. This could be then displayed in the form of Fig. 4. This "peaks tree" represents a general template for gene families evolving through WGD and fractionation-based gene loss only. The gene tree for any particular gene family would have exactly the same form, but with losses of various lineages.

4   Conclusion

We have pointed out connections between gene-tree/species-tree theory and the study of whole genome duplications in a phylogeny. The "peaks" tree should be a template for all the gene families proliferating through WGD and speciation only, where each gene family would simply require pruning of some of the branches of the tree, due to fractionation of duplicate genes. Despite the shortcomings of our Brassicaceae analysis, in the ideal case, the peaks tree itself would fill out the template completely, although no individual gene family is likely to be complete.

Our model and methodology is simplified. We have seen that λ may vary somewhat for individual lineages, and ρ is probably even more variable. Genomic processes such as chromosomal rearrangements disrupt gene order and degrade the recovery of synteny blocks and duplicate gene pairs. These issues should all be addressed in future work.

Our simplest DNA substitution model assumes equal base frequency and equal mutation rates. DNA substitution models with more parameters and rate variation among sites can be readily applied here. For example, one commonly used distance metric K_s (substitutions per synonymous sites) is typically calculated using more specific codon substitution models. The K_s distance scales linearly with time and $\log p$.

Despite the need for inference procedures focusing on the parameters λ and ρ (rather than μ and σ) jointly estimated for a compete set of similarity distributions among N genomes (rather than just one distribution), we have not yet implemented one, and have resorted to a primitive procedure of peak recognition in illustrating our model. Nevertheless, applying our concepts to six genomes in the family Brassicaceae illustrates the potential usefulness of our approach in understanding multiple WGD in a phylogenetic context.

Competing Interests
The authors declare that they have no competing interests.

Acknowledgements. Research supported in part by grants from the Natural Sciences and Engineering Research Council of Canada (NSERC). DS holds the Canada Research Chair in Mathematical Genomics.

References

1. Denoeud, F., Henriet, S., Mungpakdee, S., Aury, J.M., Da Silva, C., Brinkmann, H., Mikhaleva, J., Olsen, L.C., Jubin, C., Cañestro, C., Bouquet, J.M., Danks, G., Poulain, J., Campsteijn, C., Adamski, M., Cross, I., Yadetie, F., Muffato, M., Louis, A., Butcher, S., Tsagkogeorga, G., Konrad, A., Singh, S., Jensen, M.F., Cong, E.H., Eikeseth-Otteraa, H., Noel, B., Anthouard, V., Porcel, B.M., Kachouri-Lafond, R., Nishino, A., Ugolini, M., Chourrout, P., Nishida, H., Aasland, R., Huzurbazar, S., Westhof, E., Delsuc, F., Lehrach, H., Reinhardt, R., Weissenbach, J., Roy, S.W., Artiguenave, F., Postlethwait, J.H., Manak, J.R., Thompson, E.M., Jaillon, O., Du Pasquier, L., Boudinot, P., Liberles, D.A., Volff, J.N., Philippe, H., Lenhard, B., Crollius, H.R., Wincker, P., Chourrout, D.: Plasticity of animal genome architecture unmasked by rapid evolution of a pelagic tunicate. Science **330**, 1381–1385 (2010)
2. Haudry, A., Platts, A.E., Vello, E., Hoen, D.R., Leclercq, M., Williamson, R.J., Forczek, E., Joly-Lopez, Z., Steffen, J.G., Hazzouri, K.M., Dewar, K., Stinchcombe, J.R., Schoen, D.J., Wang, X., Schmutz, J., Town, C.D., Edger, P.P., Pires, J.C., Schumaker, K.S., Jarvis, D.E., Mandakova, T., Lysak, M.A., van den Bergh, E., Schranz, M.E., Harrison, P.M., Moses, A.M., Bureau, T.E., Wright, S.I., Blanchette, M.: An atlas of over 90,000 conserved noncoding sequences provides insight into crucifer regulatory regions. Nat. Genet. **45**, 891–898 (2013)
3. Hu, T.T., Pattyn, P., Bakker, E.G., Cao, J., Cheng, J.F., Clark, R.M., Fahlgren, N., Fawcett, J.A., Grimwood, J., Gundlach, H., Haberer, G., Hollister, J.D., Ossowski, S., Ottilar, R.P., Salamov, A.A., Schneeberger, K., Spannagl, M., Wang, X., Yang, L., Nasrallah, M.E., Bergelson, J., Carrington, J.C., Gaut, B.S., Schmutz, J., Mayer, K.F.X., Van de Peer, Y., Grigoriev, I.V., Nordborg, M., Weigel, D., Guo, Y.L.: The *Arabidopsis lyrata* genome sequence and the basis of rapid genome size change. Nat. Genet. **43**, 476–481 (2011)
4. The Arabidopsis Genome Initiative: Analysis of the genome sequence of the flowering plant *Arabidopsis thaliana*. Nature **408**, 796–815 (2000)
5. Kagale, S., Robinson, S.J., Nixon, J., Xiao, R., Huebert, T., Condie, J., Kessler, D., Clarke, W.E., Edger, P.P., Links, M.G., et al.: Polyploid evolution of the brassicaceae during the cenozoic era. Plant Cell **26**, 2777–2791 (2014)
6. Kitashiba, H., Li, F., Hirakawa, H., Kawanabe, T., Zou, Z., Hasegawa, Y., Tonosaki, K., Shirasawa, S., Fukushima, A., Yokoi, S., Takahata, Y., Kakizaki, T., Ishida, M., Okamoto, S., Sakamoto, K., Shirasawa, K., Tabata, S., Nishio, T.: Draft sequences of the radish (*Raphanus sativus* l.) genome. DNA Res. **21**(5), 481–490 (2014)
7. Liu, S., Liu, Y., Yang, X., Tong, C., Edwards, D., Parkin, I.A.P., Zhao, M., Ma, J., Yu, J., Huang, S., Wang, X., Wang, J., Lu, K., Fang, Z., Bancroft, I., Yang, T.J., Hu, Q., Wang, X., Yue, Z., Li, H., Yang, L., Wu, J., Zhou, Q., Wang, W., King, G.J., Pires, J.C., Lu, C., Wu, Z., Sampath, P., Wang, Z., Guo, H., Pan, S., Yang, L., Min, J., Zhang, D., Jin, D., Li, W., Belcram, H., Tu, J., Guan, M., Qi, C., Du, D., Li, J., Jiang, L., Batley, J., Sharpe, A.G., Park, B.S., Ruperao, P., Cheng, F., Waminal, N.E., Huang, Y., Dong, C., Wang, L., Li, J., Hu, Z., Zhuang, M., Huang, Y., Huang, J., Shi, J., Mei, D., Liu, J., Lee, T.H., Wang, J., Jin, H., Li, Z., Li, X., Zhang, J., Xiao, L., Zhou, Y., Liu, Z., Liu, X., Qin, R., Tang, X., Liu, W., Wang, Y., Zhang, Y., Lee, J., Kim, H.H., Denoeud, F., Xu, X., Liang, X., Hua, W., Wang, X., Wang, J., Chalhoub, B., Paterson, A.H.: The *Brassica oleracea* genome reveals the asymmetrical evolution of polyploid genomes. Nat. Commun. **5**, 3930 (2014). http://dx.org/10.1038/ncomms4930

8. Lyons, E., Freeling, M.: How to usefully compare homologous plant genes and chromosomes as DNA sequences. Plant J. **53**, 661–673 (2008)

9. Lyons, E., Pedersen, B., Kane, J., Freeling, M.: The value of nonmodel genomes and an example using SynMap within CoGe to dissect the hexaploidy that predates rosids. Trop. Plant Biol. **1**, 181–190 (2008)

10. McLachlan, G.J., Peel, D., Basford, K.E., Adams, P.: The Emmix software for the fitting of mixtures of normal and t-components. J. Stat. Softw. **4**(2), 1–14 (1999)

11. Ming, R., VanBuren, R., Liu, Y., Yang, M., Han, Y., Li, L.T., Zhang, Q., Kim, M.J., Schatz, M.C., Campbell, M., Li, J., Bowers, J.E., Tang, H., Lyons, E., Ferguson, A.A., Narzisi, G., Nelson, D.R., Blaby-Haas, C.E., Gschwend, A.R., Jiao, Y., Der, J.P., Zeng, F., Han, J., Min, X.J., Hudson, K.A., Singh, R., Grennan, A.K., Karpowicz, S.J., Watling, J.R., Ito, K., Robinson, S.A., Hudson, M.E., Yu, Q., Mockler, T.C., Carroll, A., Zheng, Y., Sunkar, R., Jia, R., Chen, N., Arro, J., Wai, C.M., Wafula, E., Spence, A., Han, Y., Xu, L., Zhang, J., Peery, R., Haus, M.J., Xiong, W., Walsh, J.A., Wu, J., Wang, M.L., Zhu, Y.J., Paull, R.E., Britt, A.B., Du, C., Downie, S.R., Schuler, M.A., Michael, T.P., Long, S.P., Ort, D.R., William Schopf, J., Gang, D.R., Jiang, N., Yandell, M., dePamphilis, C.W., Merchant, S.S., Paterson, A.H., Buchanan, B.B., Li, S., Shen-Miller, J.: Genome of the long-living sacred lotus (*Nelumbo nucifera* Gaertn.). Genome Biol. **14**(5), 1–11 (2013)

12. Sankoff, D., Zheng, C., Zhu, Q.: The collapse of gene complement following whole genome duplication. BMC Genomics **11**, 313–313 (2010)

13. Wang, X., Wang, H., Wang, J., Sun, R., Wu, J., Liu, S., Bai, Y., Mun, J.H., Bancroft, I., Cheng, F., Huang, S., Li, X., Hua, W., Wang, J., Wang, X., Freeling, M., Pires, J.C., Paterson, A.H., Chalhoub, B., Wang, B., Hayward, A., Sharpe, A.G., Park, B.S., Weisshaar, B., Liu, B., Li, B., Liu, B., Tong, C., Song, C., Duran, C., Peng, C., Geng, C., Koh, C., Lin, C., Edwards, D., Mu, D., Shen, D., Soumpourou, E., Li, F., Fraser, F., Conant, G., Lassalle, G., King, G.J., Bonnema, G., Tang, H., Wang, H., Belcram, H., Zhou, H., Hirakawa, H., Abe, H., Guo, H., Wang, H., Jin, H., Parkin, I.A.P., Batley, J., Kim, J.S., Just, J., Li, J., Xu, J., Deng, J., Kim, J.A., Li, J., Yu, J., Meng, J., Wang, J., Min, J., Poulain, J., Hatakeyama, K., Wu, K., Wang, L., Fang, L., Trick, M., Links, M.G., Zhao, M., Jin, M., Ramchiary, N., Drou, N., Berkman, P.J., Cai, Q., Huang, Q., Li, R., Tabata, S., Cheng, S., Zhang, S., Zhang, S., Huang, S., Sato, S., Sun, S., Kwon, S.J., Choi, S.R., Lee, T.H., Fan, W., Zhao, X., Tan, X., Xu, X., Wang, Y., Qiu, Y., Yin, Y., Li, Y., Du, Y., Liao, Y., Lim, Y., Narusaka, Y., Wang, Y., Wang, Z., Li, Z., Wang, Z., Xiong, Z., Zhang, Z.: The genome of the mesopolyploid crop species *Brassica rapa*. Nat. Genet. **43**, 1035–1039 (2011)

Biological Networks and Modelling

Relating Bisimulations with Attractors in Boolean Network Models

Daniel Figueiredo[(✉)]

Department of Mathematics, CIDMA - Center for Research
and Development in Mathematics and Applications, University of Aveiro,
Campus Universitário de Santiago, 3810-193 Aveiro, Portugal
`daniel.figueiredo@ua.pt`

Abstract. When studying a biological regulatory network, it is usual
to use boolean network models. In these models, boolean variables represent the behavior of each component of the biological system. Taking
in account that the size of these state transition models grows exponentially along with the number of components considered, it becomes
important to have tools to minimize such models. In this paper, we relate
bisimulations, which are relations used in the study of automata (general
state transition models) with attractors, which are an important feature
of biological boolean models. Hence, we support the idea that bisimulations can be important tools in the study some main features of boolean
network models. We also discuss the differences between using this approach and other well-known methodologies to study this kind of systems
and we illustrate it with some examples.

Keywords: Biological regulatory networks · Bisimulation · Minimization of models

1 Introduction

The term "biological regulatory network" refers to the regulation processes which
occur within a cell. In this environment, there are several biological components
which react with each other (for example, by chemical reactions). More generally,
the occurrence of these regulation processes within a biological system is due to
the fact that the presence of some components in the cell can either induce or
inhibit the production of some other component(s). For instance, this behavior
can be observed when some proteins interact with genes producing mRNA. In
its turn, mRNA induce the production of other proteins and so on.

To study a biological regulatory network, we must take in account that state
variables like the concentration of proteins, mRNA and other components vary

This work was supported in part by the Portuguese Foundation for Science and Technology (FCT - Fundação para a Ciência e a Tecnologia), through CIDMA - Center
for Research and Development in Mathematics and Applications, within project
UID/MAT/04106/2013. The author also acknowledges the support of FCT via the
Ph.D. scholarship PD/BD/114186/2016.

© Springer International Publishing Switzerland 2016
M. Botón-Fernández et al. (Eds.): AlCoB 2016, LNBI 9702, pp. 17–25, 2016.
DOI: 10.1007/978-3-319-38827-4_2

in a continuous form. Indeed, one of the most precise kind of models used in this field are those which describe the dynamics of a biological regulatory network by an ordinary differential equations system (see [5]) that only admits continuous state variables. Usually, these models use sigmoid functions to describe a positive/negative regulation of a component over another (*i.e.*, to describe that one component induces/inhibits the production of another). The sigmoid functions which are more often used to describe a positive regulation are the so called "Hill functions" and depend on parameters θ and n. In this way, it is not difficult to see that this kind of models admit non linear equations and, therefore, the resulting ordinary differential equations system is not trivially solved by analytic methods. Thus, other (more simple) kinds of models are often used in order to proceed with a preliminary study of the biological system. In this context, the boolean networks are really useful.

There are many variants of boolean network models, however, the basic idea of all them is to approach each state variable of the system by a boolean variable [9]. In this way, it is assigned either the value "1" or "0" to indicate that some component is present or absent, respectively. Then, for each component i we define a boolean variable x_i and consider a threshold θ_i. If the concentration of the component i is above θ_i we define x_i as "1" (present) and otherwise we define it as "0" (absent).

In a boolean network, a state is a vector $(x_1, ..., x_n)$ such that each x_i is the boolean variable associated to the component i. There also are some variants of these boolean approaches which admit more than one threshold associated to each component: θ_i^j. In this case, it is possible to obtain several levels of expressibility which are codified using several boolean variables x_i^j instead of only x_i. Still, there are other variants as asynchronous boolean networks (see [2]) which we will more carefully describe in this paper. A boolean network model which represents the dynamics of a biological system is a digraph in which each vertex is a vector composed of "0"s and "1"s (which relates to a possible configuration of the biological system); and each edge relates to a possible transition between states. We then have a state transition model (automata).

These biological models and their variants are widely used because they are simple and some features of the original system can be identified by studying these boolean models. One of the most studied features in biological regulatory networks is the existence of *steady states*. By steady state we refer to the values of the concentration of a cell's components where the system stabilizes. In particular, in models which use an ODE system, a steady state corresponds to those states in which the evolution of the system is null, *i.e.*, corresponds to set all differential equations to zero. When a steady state exists, it can be either stable (if little perturbations do not cause the system to evolve into a state far away from the initial steady state) or unstable (otherwise). Thus, the study of these characteristics is an important topic in the field of biological systems. Because of this, discrete models as boolean networks are often used because it is well-known that steady states are signaled by terminals in asynchronous boolean networks. Then, it becomes worth to use these models to proceed with a preliminary study.

We point out that in a biological context, the concept *attractor* is more often used than "terminal". Therefore, we use the term "attractor" instead of "terminal" when we refer to this concept in biological boolean networks.

It is not difficult to see that the number of states of these models grows exponentially with the number of components of the system. For example, a model which considers 10 components admits 2^{10} states. Because of this, and since the most of the biological models admits much different components (usually, much more than 10), it becomes important to both develop tools to minimize these boolean network models and new ideas to find features like attractor with few computational cost. In order to do this, we propose to take into account the ideas already used in automata theory. Although we do not present any new algorithm, this work paves the way to new approaches to this problem.

In this paper, we apply the concept of bisimulation to propose a new method to preliminarily study these biological systems. Bisimulations are already used in several minimization processes. Furthermore, the possibility of combining bisimulation with modal logic which admit modalities (see [1]) turns out that it is a powerful tool to study state transition systems. The usage of modal logic is possible due to the possibility of interpreting biological boolean network as Kripke models. However, in this paper, we will only propose bisimulations to develop new minimization processes which allow us to find the attractors of boolean networks and we do not consider any background logic. Thus, given a digraph (V, E), we say that $S \subseteq V \times V$ is a bisimulation if S is not empty, and if it is an equivalence relation such that:

- If $(v, w) \in S$ and $(v, v') \in E$ then there exists $w' \in V$ such that $(w, w') \in E$ and $(v', w') \in S$.
- If $(v, w) \in S$ and $(w, w') \in E$ then there exists $v' \in V$ such that $(v, v') \in E$ and $(v', w') \in S$.

Outline. We begin by presenting some definitions and a theorem that relates attractors with bisimulations. Then, we enhance the difference between minimizing boolean networks using bisimulation and other known methodologies used in the study of such systems. Finally, we present some conclusions and directions to follow.

2   Bisimulations and Attractors

The dynamics within a cell are guided by several components: proteins, RNA, genes, ribosomes, *etc*... Each of these components induces or inhibits the production/activation of some of the other ones. Thus, it is very difficult to understand a biological regulatory network and, usually, only some main features are studied. As referred, one of these features are the steady states. Steady states are related to the modes of operating of a cell. For instance, in [4], a model for *E. coli* with two steady states is presented: one is related to a configuration in which the organism metabolizes sugar, grows and replicate itself; and the other relates to a configuration in which the organism does not metabolize sugar, does not grow and does not replicate itself.

There exists results which shows that the study of asynchronous boolean models can be used to identify steady states (see [9]). In practice, each terminal represents a steady state. We follow with a formal definition of terminal.

Definition 1. Let (V, E) be a graph.

We say that $v \in V$ has a transition to $w \in V$ and we write $v \rightarrow w$, if $(v, w) \in E$. We write $v \not\rightarrow w$ otherwise.

We say that there is a *path* from v to w if there exist $v_1, ..., v_n \in V$ such that $v \rightarrow v_1, v_1 \rightarrow v_2, ..., v_n \rightarrow w$.

A Strongly Connected Component (SCC) is a subset A of V such that there is a path between any two element of A.

A set A is a terminal if it is a SCC and $\nexists\, a \in A, v \in V \backslash A$: $a \rightarrow v$.

We point out that, since the biological boolean models represent a finite number of components of a cell, in this section, we assume that all the considered digraphs are finite, *i.e.*, for any digraph (V, E), $|V| < \infty$.

To find the attractors of digraphs, several methodologies can be used. In [7,10] some methods are presented. However, here, we present some new ideas that can lead to a new approach on this theme, based in bisimulations. We thus follow with a definition of a particular class of bisimulations and we present and prove a theorem that relates this class of bisimulations with attractors.

Definition 2. Let (V, E) be a graph.

We say that $\mathcal{B} \subseteq V \times V$ is a complete bisimulation if it is a bisimulation and there exists $B \subseteq V$ such that $\mathcal{B} = B \times B$ (any two elements of B are related).

We say that a complete bisimulation $\mathcal{B}$ is minimal if there is not any other complete bisimulation $\mathcal{B}'$ such that $\mathcal{B}' \subsetneq \mathcal{B}$.

Lemma 1. *Let (V, E) be a graph, $B \subseteq V$ and $\mathcal{B} = B \times B$ a minimal complete bisimulation. For any $A \subsetneq B$, $\exists\, a \in A, v \in B \backslash A$ such that $a \rightarrow v$.*

Proof. Let us assume that there exists $A \subsetneq B$ such that, for any $a \in A, v \in B \backslash A$, $a \not\rightarrow v$.

In this case, we can easily verify that $A \times A$ is an equivalence relation since all states of A are related. By hypothesis, for any $(a, a') \in A \times A \subseteq \mathcal{B}$ such that $a \rightarrow b$, there exists some b' which verifies $a' \rightarrow b'$ and $(b, b') \in \mathcal{B}$. Since for any $a \in A, v \in B \backslash A$, $a \not\rightarrow v$, we can conclude that $b, b' \in A$ and, therefore, $(b, b') \in A \times A$. Thus, $A \times A$ is a complete bisimulation and this contradicts the minimality of $\mathcal{B}$. □

Theorem 1. *Let (V, E) be a graph. $\mathcal{B} = B \times B \subseteq V \times V$ is a minimal complete bisimulation $\Leftrightarrow B$ is a terminal of V.*

Proof.

"$\Rightarrow$"

We start by proving that if $\mathcal{B}$ is a minimal complete bisimulation, then there exists a path between any two elements of B. We prove that B is a terminal afterwards.

We consider $u, v \in B$. By Lemma 1, we know that there is a transition $u \to u_1$ from $\{u\}$ to $B\backslash\{u\}$. If $u_1 = v$, we are done. Otherwise, using Lemma 1 again, we know that there is a transition from $\{u, u_1\}$ to some $u_2 \in B\backslash\{u, u_1\}$. Here, either $u \to u_2$ or $u_1 \to u_2$. In any case, there is a path from u to u_2. Again, if $u_2 = v$ we are done. Otherwise, we can continue to apply this procedure till find a path between u and v (this procedure will end in finite time since we are only considering finite graphs). As u and v were arbitrary, we can conclude that B is a SCC.

Finally, if $u \in B$ and $u \to v$, then $(u, u) \in \mathcal{B}$ and, by definition of complete bisimulation, $(v, v) \in \mathcal{B}$. Then $v \in B$ and, thus, B is a terminal.

"$\Leftarrow$"

We now assume that B is a terminal of V. We can easily see that $\mathcal{B} = B \times B$ is a equivalence relation since all states are related. We consider $(u, v) \in \mathcal{B}$ and $u \to u'$. Since B is a terminal, $u' \in B$ and $\exists v' \in B$ such that $v \to v'$. Furthermore, $(v, v') \in \mathcal{B}$, by definition and, therefore, $\mathcal{B}$ is a complete bisimulation.

Let us assume that $\mathcal{B}$ is not minimal, $i.e.$, there is a complete bisimulation $\mathcal{A} := A \times A \subsetneq \mathcal{B}$. Since B is terminal, it is possible to find a path from any $a \in A$ for any $b \in B\backslash A$. Thus, $\exists a' \in A, b' \in B\backslash A$ such that $a' \to b'$. But this contradicts the fact of $\mathcal{A}$ being bisimulation because $b' \notin A$. □

This theorem can help us to develop a new minimization methodology which preserves the attractors without even know them. The general idea is to find a bisimulation $\mathcal{B}$ such that any complete bisimulation contained in $\mathcal{B}$ is minimal. Thus, we can compute the "quotient digraph" in order to obtain a minimized model in which the states of the attractors may be "clustered". Nevertheless, these attractors are individually preserved. We follow with two examples in order better understand how the minimization via bisimulation is made.

In this example, we consider two asynchronous boolean networks models. In this kind of boolean network, the directed edges representing the transitions between states are defined according to some boolean equations. However, we can only update the value of one variable at each time. To simplify, we do not consider any loops.

For the first example, we pick a purely theoretical model which is presented in Fig. 1. This example is presented in order to further distinguish two methodologies. In this figure, the attractor of the model is enhanced by an orange box. Although this model is theoretical, it could result from a system comporting three components (a, b and c) whose state transition is computed by updating the value of a single component at each time and according to the following boolean equations:

$$\begin{cases} a := a \vee b \\ b := a \vee (b \oplus c) \\ c := (a \oplus \neg b) \vee c \end{cases}$$

In these equations, $\oplus$ is the XOR boolean operator.

In order to minimize this model, we can find a bisimulation $\mathcal{B} = \{(000, 000), (001, 001), (011, 011), (111, 111), (010, 010), (110, 110), (100, 100), (101, 101),$

Fig. 1. Model with an attractor composed of a single state.

$(001, 011), (011, 001), (110, 101), (101, 110), (010, 100), (100, 010)\}$. It is not difficult to verify that all complete bisimulation contained in this bisimulation are
minimal. Thus, we can construct the "quotient digraph" by clustering the states
in the same equivalence class. The transitions of the "quotient digraph" are
introduced by the following rule: "If $a \to b$, then, $[a] \to [b]$ (where $[a]$ and $[b]$ are
the equivalence classes of a and b, respectively)". This quotient digraph is then
presented in Fig. 2.

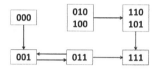

Fig. 2. Quotient digraph of the model in Fig. 1.

We now consider a real example. In [3], it is presented a biological system that
regulates the circadian rhythm in a cyanobacteria, *i.e.* this system models the
biological processes (which are periodic and whose period is 24 h) that regulates
the perception of a day-cycle by an organism. In [3], this system is studied with
a asynchronous boolean network and the attractors of the resulting network are
found and it is studied the robustness of this model. The asynchronous boolean
network used is defined by the boolean equations which follow:

$$
\begin{cases}
a := \neg s \\
s := ts \\
t := a \\
ts := t \wedge a
\end{cases}
$$

This model describes the dynamics of the three phosphorylated forms of Kai
C (using boolean variables t, ts, and s) and the protein Kai A (with a boolean
variable a). These four form are responsible for the core of the cyanobacterial circadian clock. The referred system is modeled by the presented boolean functions
and the resulting asynchronous boolean network, whose attractor is enhanced
by an orange box, is shown in Fig. 3.

As before, we can find a bisimulation such that all complete bisimulations contained on it are minimal and compute the "quotient digraph".
Hence, if we consider the following bisimulation $\mathcal{B} = \{(a, b) : a, b \in$

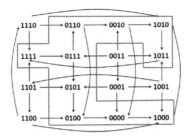

Fig. 3. Model of the circadian rhythm in a cyanobacteria.

$\{0110, 0010, 1010, 1111, 0111, 1011, 0101, 0100, 0000, 1000\}\} \cup \{(a,a) : a \in \{1110, 1101, 1100, 0011, 0001, 1001\}\}$, the resultant quotient digraph is the one which is presented in Fig. 4.

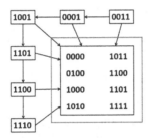

Fig. 4. Minimized model for the circadian rhythm in a cyanobacteria.

In both cases, we can see that we obtain a minimized model in which the attractors are preserved. This is due to the fact that all complete bisimulations contained in the bisimulation used are minimal.

3 Comparing Bisimulation with Other Reducing Methods

In this section, we compare this method of minimization with other methodologies which are commonly used in this field. We do this in order to distinguish our minimizing method from those and we point out some advantages (and disadvantages) of our method when comparing to other approaches.

Firstly, we compare our quotient digraph with the hierarchical representations. This kind of representations sort the states of a model according to their distance to the attractors. This approach is widely used in the study of systems. However, the main disadvantage is that one must know *a priori* which are the attractors of the model in order to obtain a hierarchical representation. It is not difficult to see that our method can only get together two states whose distance to the attractors is the same (since transitions are someway preserved). However it only clusters states whose behavior is similar. To illustrate this we call the minimized model in Fig. 4. In this example, if we consider any state which is not part of the attractor, we can see that its distance to the attractor is "1". Despite this fact, we can see that any two of these states are not clustered because each one of them presents a different behavior. Actually it may be important to study these differences in their behavior. Actually, they allow us to discover the longest possible "path" to the attractor.

Another widely used method to minimize boolean networks and, then, search for attractors is clustering the SCCs. This allows us to minimize a model and still preserve the attractors. This is a well-known idea and several other methods to find attractor where developed after it (for instance, see [10]). On one hand, the method which we present can clusters sets of states which are not in the same SCC and, therefore, it clusters states which would not be clustered when we cluster the SCC's. For instance, recalling the example in Fig. 2, we can see that the states 010 and 100 were clustered and were not in the same SCC; on the other hand, constructing the quotient digraph, it can happen that we do not cluster all SCC's. For example, the quotient digraph presented in Fig. 2 still has a SCC. This is due to the fact that our method can only cluster SCCs which are terminals. We can see this because, for any SCC which is not a terminal, there exists some state in the SCC that admits a transition for a state out of that SCC. Therefore, it may be impossible to find a complete bisimulation to cluster all its states.

Finally, we point out a last important feature of bisimulations. Since we are dealing with discrete state transition models (automata), it can be useful to use modal logic to reason about such models. Hence, it could be useful to obtain minimization processes which guarantee that all states in a cluster verify the same modal formulas. Indeed, due to their definition, bisimulations are suitable to be used with modal logic. More information about this can be found in [1].

4   Conclusions

Bisimulations can be used to obtain to minimize biological boolean models and, guaranteeing some conditions, the methodology we presented preserves the attractors. Although we present only the main ideas and some examples of the application of this method, it can provide the basis for a new minimization algorithm. Actually, in future, we are planing to develop this complete algorithm which applies these ideas to minimize biological boolean models.

We also evaluated the convenience of using this minimization methodology when compared with other methods already used. It provides a new way of looking at biological models and it can be useful in their study. When comparing with other methods, it has both some advantages and disadvantages. However, as seen, since it preserves the attractors and, moreover, it can be combined with a modal logic, we believe that this approach is worth.

In future, we also plan to study how can modal logic fit in these biological boolean models and, if possible, to find an axiomatization of such systems which would allow us to formally prove diverse properties of them. Actually, in continuous models which use ODEs, it was applied a dynamic logic (which integrates first-order features) proposed by A. Platzer – Differential Dynamic Logic (see [8]) – to formally reason about them. Some initial work can be found in [6]. We believe that is possible to obtain a similar results in discrete models and, in particular, in boolean networks.

References

1. Blackburn, P., De Rijke, M., Venema, Y.: Modal Logic: Graph. Darst, vol. 53. Cambridge University Press, Cambridge (2002)
2. Chaves, M.: Predictive analysis of dynamical systems: combining discrete and continuous formalisms. Ph.D. thesis, Gipsa-lab (2013)
3. Chaves, M., Preto, M.: Hierarchy of models: from qualitative to quantitative analysis of circadian rhythms in cyanobacteria. Chaos Interdisc. J. Nonlinear Sci. $\mathbf{23}(2)$, 025113 (2013)
4. Chaves, M., Tournier, L.: Predicting the asymptotic dynamics of large biological networks by interconnections of boolean modules. In: 2011 50th IEEE Conference on Decision and Control and European Control Conference (CDC-ECC), pp. 3026–3031. IEEE (2011)
5. De Jong, H.: Modeling and simulation of genetic regulatory systems: a literature review. J. Comput. Biol. $\mathbf{9}(1)$, 67–103 (2002)
6. Figueiredo, D.: Differential dynamic logic and applications. Master's thesis, University of Aveiro (2015)
7. Naldi, A., Remy, E., Thieffry, D., Chaouiya, C.: A reduction of logical regulatory graphs preserving essential dynamical properties. In: Degano, P., Gorrieri, R. (eds.) CMSB 2009. LNCS, vol. 5688, pp. 266–280. Springer, Heidelberg (2009)
8. Platzer, A.: Logical Analysis of Hybrid Systems: Proving Theorems for Complex Dynamics. Springer, Heidelberg (2010)
9. Thomas, R., Kaufman, M.: Multistationarity, the basis of cell differentiation and memory. II. Logical analysis of regulatory networks in terms of feedback circuits. Chaos Interdisc. J. Nonlinear Sci. $\mathbf{11}(1)$, 180–195 (2001)
10. Tournier, L., Chaves, M.: Interconnection of asynchronous boolean networks, asymptotic and transient dynamics. Automatica $\mathbf{49}(4)$, 884–893 (2013)

Neural Networks Simulation of Feeding Adaptations of Daphnia

Tibor Kmet[1]([⊠]) and Maria Kmetova[2]

[1] Department of Informatics, Constantine the Philosopher University,
Tr. A. Hlinku 1, 949 74 Nitra, Slovakia
tkmet@ukf.sk
[2] Department of Mathematics, Constantine the Philosopher University,
Tr. A. Hlinku 1, 949 74 Nitra, Slovakia
mkmetova@ukf.sk
http://www.ukf.sk

Abstract. In this article we present a neutral network based optimal control synthesis for solving distributed optimal control problems for systems governed by parabolic differential equations with control and state constraints and discrete time delay. The optimal control problem is transcribed into nonlinear programming problem which is implemented with feed forward adaptive critic neural network to find optimal control and optimal trajectory. The developed simulation method is demonstrated on the optimal control problem of feeding adaptation of Daphnia model with diffusion and discrete time delay of nutrient uptake. Results show that adaptive critic based systematic approaches are promising in obtaining the optimal distributed control with discrete time delays in state and control variables subject to control and state constraints.

Keywords: Distributed control problem with discrete time delays · State and control constraints · Feed-forward neural network · Adaptive critic synthesis · Numerical examples

1 Introduction

We consider an optimal distributed control problem for systems governed by a parabolic differential equations, with control and state constraints and discrete time delay. The problem is motivated by better understanding of real world systems eventually with the purpose of being able to influence these systems in a desired way. The scope of this paper is the study of discretization/optimization methods using neural networks generating controls. We pursue the one-shot multigrid strategy as proposed in [1]. A one-shot multigrid algorithm means to solve the optimality system for the state, the adjoint and the control variables in the multigrid process in parallel. The finite element approximation plays an important role in the numerical treatment of optimal control problems. This approach has been extensively studied in the papers e.g. [3,4,9–11] for parabolic optimal control problems. Through discretization the optimal control problem

© Springer International Publishing Switzerland 2016
M. Botón-Fernández et al. (Eds.): AlCoB 2016, LNBI 9702, pp. 26–37, 2016.
DOI: 10.1007/978-3-319-38827-4_3

is transcribed into a finite-dimensional nonlinear programming problem (NLP-problem). The basic idea of these methods is to apply nonlinear programming techniques to the resulting finite dimensional optimization problem [2,5,11]. Then neural networks are used as a universal function approximation to solve finite dimensional optimization problems forward in time with "adaptive critic designs" [12,15]. For the neural network, a feed forward neural network with one hidden layer, a steepest descent error backpropagation rule, a hyperbolic tangent sigmoid transfer function and a linear transfer function were used. The presented paper extends adaptive critic neural network architecture proposed by [7] to the optimal distributed control problem for systems governed by a parabolic differential equations with control and state constraints and discrete time delay. This paper is organized as follows. In Sect. 2, optimal distributed control problems with delays in state and control variables subject to control and state constraints are introduced. We summarize the necessary optimality conditions and give a short overview of the basic results including the iterative numerical methods. In Sect. 3, we discuss the discretization methods for the given optimal control problem and formulate the resulting nonlinear programming problems. Section 4 presents a short description of adaptive critic neural network synthesis for the optimal control problem with delays in state and control variables subject to control and state constraints. We also present a new algorithm to solve optimal control problems. In Sect. 5, we present a description of feeding adaptations of Daphnia. We apply the new proposed methods to the model presented to compare short-term and long-term strategies of nutrients uptake by Daphnia. Numerical results are also given. Conclusions are being presented in Sect. 6.

2   The Optimal Control Problem

We consider the nonlinear control problem governed by parabolic equations with delays in state and control variables subject to control and state constraints. Let $x(p,t) \in R^n$ and $u(p,t) \in R^m$ denote the state and control variable, respectively in a given space-time domain $Q = [a,b] \times [t_0, t_f]$. The optimal control problem is to minimize

$$J(u) = \int_a^b g(x(p, t_f)) dp \qquad (1)$$

$$+ \int_a^b \int_{t_0}^{t_f} f_0(x(p,t), x(p, t - \tau_x), u(p,t), u(p, t - \tau_u)) dt dp,$$

subject to

$$\frac{\partial x(p,t)}{\partial t} = D\frac{\partial^2 x(p,t)}{\partial p^2} + f(x(p,t), x(p, t - \tau_x), u(p,t), u(p, t - \tau_u)), \quad (2)$$

$$\frac{\partial x(a,t)}{\partial p} = \frac{\partial x(b,t)}{\partial p} = 0, \quad t \in [t_0, t_f],$$

$$x(p,t) = \phi_s(p,t), \quad u(p,t) = \phi_c(p,t), \quad p \in [a,b], \quad t \in [t_0 - \tau_u, t_0],$$

$$\psi(x(p, t_f)) = 0, \quad c(x(p,t), u(p,t)) \leq 0, \quad p \in [a,b], \quad t \in [t_0, t_f],$$

where $\tau_x \geq 0$ and $\tau_u \geq 0$ are discrete time delay in the state and control variable, respectively. The functions $g : R^n \to R$, $f_0 : R^{2(n+m)} \to R$, $f : R^{2(n+m)} \to R^n$, $c : R^{n+m} \to R^q$ and $\psi : R^n \to R^r$, $0 \leq r \leq n$ are assumed to be sufficiently smooth on appropriate open sets, and the initial conditions $\phi_s(p, t)$, $\phi_c(p, t)$ are continuous functions. The theory of necessary conditions for the optimal control problem of form (1) is well developed, see e.g. [5, 11]. We introduce an additional state variable $x_0(p, t) = \int_0^t f_0(x(s), x(s - \tau_x), u(s), u(s - \tau_u)) ds$. Then the augmented Hamiltonian function for problem (1) is

$$\mathcal{H}(x, x_{\tau_x}, u, u_{\tau_u}, \lambda, \mu) = \sum_{j=0}^n \lambda_j f_j(x, x_{\tau_x}, u, u_{\tau_u}) + \sum_{j=0}^q \mu_j c_j(x, u),$$

where $\lambda \in R^{n+1}$ is the adjoint variable and $\mu \in R^q$ is a multiplier associated to the inequality constraints. Assume that τ_x, $\tau_u \geq 0$, $(\tau_x, \tau_u) \neq (0, 0)$ and $\frac{\tau_x}{\tau_u} \in \mathbb{Q}$ for $\tau_u > 0$ or $\frac{\tau_u}{\tau_x} \in \mathbb{Q}$ for $\tau_x > 0$. Let $(\hat{x}, \hat{u})$ be an optimal solution for (1). Then the necessary optimality condition for (1) implies [5] that there exist a piecewise continuous and piecewise continuously differentiable adjoint function $\lambda : Q \to R^{n+1}$, a piecewise continuous multiplier function $\mu : Q \to R^q$, $\hat{\mu}(p, t) \geq 0$ and a multiplier $\sigma \in R^r$ satisfying

$$\frac{\partial \lambda}{\partial t} = D \frac{\partial^2 \lambda}{\partial p^2} - \frac{\partial \mathcal{H}}{\partial x} (\hat{x}, \hat{x}_{\tau_x}, \hat{u}, \hat{u}_{\tau_u}, \lambda, \mu)$$
$$- \chi_{[t_0, t_f - \tau_x]} \frac{\partial \mathcal{H}}{\partial x_{\tau_x}} (\hat{x}_{+\tau_x}, \hat{x}, \hat{u}_{+\tau_x}, \hat{u}_{\tau_u + \tau_x}, \lambda_{+\tau_x}, \mu_{+\tau_x}), \qquad (3)$$

$$\lambda(p, t_f) = g_x(\hat{x}(p, t_f)) + \sigma \psi_x(\hat{x}(p, t_f)), \frac{\partial \lambda(a, t)}{\partial p} = \frac{\partial \lambda(b, t)}{\partial p} = 0, \qquad (4)$$

$$0 = -\frac{\partial \mathcal{H}}{\partial u} (\hat{x}, \hat{x}_{\tau_x}, \hat{u}, \hat{u}_{\tau_u}, \lambda, \mu)$$
$$- \chi_{[t_0, t_f - \tau_u]} \frac{\partial \mathcal{H}}{\partial u_{\tau_u}} (\hat{x}_{+\tau_u}, \hat{x}_{\tau_x + \tau_u}, \hat{u}_{+\tau_u}, \hat{u}, \lambda_{+\tau_u}, \mu_{+\tau_u}). \qquad (5)$$

Furthermore, the complementary conditions hold, i.e. in $p \in [a, b]$, $t \in [t_0, t_f]$, $\mu(p, t) \geq 0$, $c(x(p, t), u(p, t)) \leq 0$ and $\mu(p, t) c(x(p, t), u(p, t)) = 0$. Herein, the subscript x, x_{τ_x}, u and u_{τ_u} denotes the partial derivative with respect to x, x_{τ_x}, u and u_{τ_u}, respectively and $x_{+\tau_x} = x(p, t+\tau_x)$, $x_{\tau_x + \tau_u} = x(p, t - \tau_x + \tau_u)$.

3   Discretization of the Optimal Control Problem

The purpose of this section is to develop discretization techniques by which the distributed control problem (1) are transformed into a nonlinear programming problem (NLP-problem) [1, 3, 11]. We assume that $\tau_u = l \frac{\tau_x}{k}$ with $l, k \in \mathbb{N}$.

Defining $h_{max} = \frac{T_x}{k}$ gives the maximum interval length for an elementary transformation interval that satisfies $\frac{T_x}{h_{max}} = k \in \mathbb{N}$ and $\frac{T_u}{h_{max}} = l \in \mathbb{N}$. The minimum grid point number for an equidistant discretization mesh $N_{min} = \frac{t_f - t_0}{h_{max}}$. Choose a natural number $K \in \mathbb{N}$ and set $N = K N_{min}$. Let $t_j \in \langle t_0, t_f \rangle$, $j = 0, \ldots, N$, be an equidistant mesh point with $t_j = t_0 + jh_t, i = 0, \ldots, N$, where $h_t = \frac{b-a}{N}$ is a time step and $t_f = Nh + t_0$. Assume that the rectangle $R = \{(p, t) : a \leq p \leq b, t_0 \leq t \leq t_f\}$ is subdivided into N by M rectangles with sides h_t and $h_s = \frac{b-a}{M}$. Start at the bottom row, where $t = t_0$, and the solution is $x(p_i, t_0) = \phi_s(p_i, t_0)$. A method for computing the approximations to $x(p, t)$ at grid points in successive rows $\{x(p_i, t_j) : i = 0, 2, \ldots, N, \ j = 0, 2, \ldots, M\}$ will be developed. The difference formulae for $x_t(p, t)$ and $x_{pp}(p, t)$ are

$$x_t(p, t) \approx \frac{x(p, t + h_t) - x(p, t)}{h_t}, \quad x_p(p, t) \approx \frac{x(p + h_s, t) - x(p, t)}{h_s} \tag{6}$$

and

$$x_{pp}(p, t) \approx \frac{x(p - h_s, t) - 2x(p, t) + x(p + h_s, t)}{h_s^2}. \tag{7}$$

Use the approximation $x_{i,j}$ in Eqs. (6) and (7), which are in turn substituted into Eq. (2), to obtain

$$\frac{x_{i,j+1} - x_{i,j}}{h_t} = D \frac{x_{i-1,j} - 2x_{i,j} + x_{i+1,j}}{h_s^2} + f_{i,j}. \tag{8}$$

Equation (8) is applied to create the $(j + 1)$th row across the grid, assuming that approximations in the jth row are known. Let the vectors $x_{ij} \in R^n$, $u_{ij} \in R^m$, $i = 0, \ldots, N$, $j = 0, \ldots, M$, be an approximation of the state variable and control variable $x(p_i, t_j)$, $u(p_i, t_j)$, respectively at the mesh point (p_i, t_j). $z := ((x_{ij}), (u_{ij}), i = 0, \ldots, N, j = 0, \ldots, M) \in R^{N_s}$, $N_s = (n+m)NM$, the optimal control problem is replaced by the following discretized control problem in the form of nonlinear programming problem with inequality constraints: Minimize

$$J(z) = h_s h_t \sum_{(i,j)} f_0(x_{ij}, x_{\tau_x ij}, u_{ij}, u_{\tau_u ij}) + h_s \sum_{(i)} g(x_{iM}) \tag{9}$$

subject to

$$x_{i,j+1} = x_{ij} + h_t D \frac{x_{i-1,j} - 2x_{ij} + x_{i+1,j}}{h_s^2} + h_t f_{ij}, \ x_{0j} = x_{1j}, \ x_{Nj} = x_{N-1j}, \tag{10}$$

$$x_{i,-j} = \phi_x(p_i, t_0 - jh), \quad j = k, \ldots, 0, \quad u_{i,-j} = \phi_u(p_i, t_0 - jh), \quad j = l, \ldots, 0,$$

$$\psi(x_{i,N}) = 0, \quad c(x_{ij}, u_{ij}) \leq 0, \ i = 0, \ldots, N, \ j = 0, \ldots, M - 1.$$

In a discrete-time formulation we want to find an admissible control which minimizes objective function (9). Let us introduce the Lagrangian function for the nonlinear optimization problem (9):

$$\mathcal{L}(z,\lambda,\sigma,\mu) = h_s h_t \sum_{(i,j)} f_0(x_{ij}, x_{\tau_x ij}, u_{ij}, u_{\tau_u ij}) + h_s \sum_{(i)} g(x_{iM})$$

$$+ \sum_{(i,j)} \lambda_{i,j+1}\left(-x_{i,j+1} + x_{ij} + D\frac{x_{i-1,j} - 2x_{ij} + x_{i+1,j}}{h_s^2} + h_t f_{ij}\right)$$

$$+ \sum_{(i,j)} \mu_{ij} c(x_{ij}, u_{ij}) + \sum_{(i)} \sigma_i \psi(x_{iN})$$

$$+ \sum_{(j)} \lambda_{0,j}\frac{x_{0,j+1} - \lambda_{0,j}}{h_s} + \sum_{(j)} \lambda_{N,j}\frac{x_{N,j+1} - \lambda_{N,j}}{h_s}. \tag{11}$$

The first order optimality conditions of Karush-Kuhn-Tucker for the problem (9) are:

$$0 = \mathcal{L}_{x_{ij}}(z,\lambda,\sigma,\mu) = \lambda_{i,j+1} - \lambda_{ij} + h_t D\frac{\lambda_{i-1,j} - 2\lambda_{ij} + \lambda_{i+1,j}}{h_s^2}$$

$$+ h_t \lambda_{i,j+1} f_{x_{ij}}(x_{ij}, x_{i,j-k}, u_{ij}, u_{i,j-l}) \tag{12}$$

$$+ h_t \lambda_{i,j+k+1} f_{x_{ij\tau_x}}(x_{i,j+k}, x_{ij}, u_{i,j+k}, u_{i,j-l+k})$$

$$+ \mu_{ij} c_{x_{ij}}(x_{ij}, u_{ij}),$$

$$j = 0, \ldots, M-k-1,$$

$$0 = \mathcal{L}_{x_{ij}}(z,\lambda,\sigma,\mu) = \lambda_{i,j+1} - \lambda_{ij} + h_t D\frac{\lambda_{i-1,j} - 2\lambda_{ij} + \lambda_{i+1,j}}{h_s^2}$$

$$+ h_t \lambda_{i,j+1} f_{x_{ij}}(x_{ij}, x_{i,j-k}, u_{ij}, u_{i,j-l}),$$

$$j = M-k, \ldots, M-1,$$

$$0 = \mathcal{L}_{x_{iM}}(z,\lambda,\sigma,\mu) = g_{x_{im}}(x_{im}) + \sigma_i \psi_{x_{iM}}(x_{iM}) - \lambda_{iM}, \lambda_{0j}$$

$$= \lambda_{1j}, \lambda_{Nj} = \lambda_{N-1,j} \tag{13}$$

$$0 = \mathcal{L}_{u_{ij}}(z,\lambda,\sigma,\mu) = h_t \lambda_{i,j+1} f_{u_{ij}}(x_{ij}, x_{i,j-k}, u_{ij}, u_{i,j-l})$$

$$+ h_t \lambda_{i,j+l+1} f_{u_{ij\tau_u}}(x_{i,j+l}, x_{i,j-k+l}, u_{i,j+l}, u_{i,j})$$

$$+ \mu_{ij} c_{x_{ij}}(x_{ij}, u_{ij}),$$

$$j = 0, \ldots, M-l-1 \tag{14}$$

$$0 = \mathcal{L}_{u_{ij}}(z,\lambda,\sigma,\mu) = h_t \lambda_{i,j+1} f_{u_{ij}}(x_{ij}, x_{i,j-k}, u_{ij}, u_{i,j-l}) + \mu_{ij} c_{x_{ij}}(x_{ij}, u_{ij}),$$

$$j = M-l, \ldots, N-1.$$

Equaitons (12)–(14) represent the discrete version of the necessary condition (3)–(5) for optimal control problem (1).

4 Adaptive Critic Neural Network for an Optimal Control Problem with Control and State Constraints

It is well known that a neural network can be used to approximate the smooth time-invariant functions [6]. Experience has shown that optimization of functionals over admissible sets of functions made up of linear combinations of relatively few basis functions with a simple structure and depending nonlinearly on a set of "inner" parameters e.g., feedforward neural networks with one hidden layer and linear output activation units often provides surprisingly good suboptimal solutions [6].

Let $x = [x_1, \ldots, x_n]'$ and $y = [y_1, \ldots, y_m]'$ be the input and output vectors of the network, respectively. Let $V = [v_1, \ldots, v_r]'$ be the matrix of synaptic weights between the input nodes and the hidden units, where $v_k = [v_{k0}, v_{k1} \ldots, v_{kn}]$; v_{k0} is the bias of the kth hidden unit, and v_{ki} is the weight that connects the ith input node to the kth hidden unit. Let also $W = [w_1, \ldots, w_m]'$ be the matrix of synaptic weights between the hidden and output units, where $w_j = [w_{j0}, w_{j1} \ldots, w_{jr}]$; w_{j0} is the bias of the jth output unit, and w_{jk} is the weight that connects the kth hidden unit to the jth output unit. The response of the kth hidden unit is

---

Algorithm 1. Algorithm to solve the optimal control problem.

Input: Choose t_0, t_f, a, b, N, M - number of steps, time and space steps
h_t, h_s ε_a, ε_c - stopping tolerance for action and critic neural networks, respectively, $x_{i,-j} = \phi_s(p_i, t_0 - jh_t)$, $j = k, \ldots, 0$,
$u_{i,-j} = \phi_c(i, t_0 - jh_t)$, $j = l, \ldots, 0$ -initial values.

Output: Set of final approximate optimal control $\hat{u}(p_i, t_0 + jh_t) = \hat{u}_{ij}$ and optimal trajectory $\hat{x}(p_i, t_0 + (j+1)h_t) = \hat{x}_{i,j+1}$, $j = 0, \ldots, M - 1$, respectively

1   Set the initial weight $\mathbb{W}^a = (V^a, W^a)$, $\mathbb{W}^c = (V^c, W^c)$
    for $j \leftarrow 0$ **to** $M - 1$ **do**
2   **for** $i \leftarrow 1$ **to** $N - 1$ **do**
3   **while** $err_a \geq \epsilon_a$ **and** $err_c \geq \epsilon_c$ **do**
4   **for** $s \leftarrow 0$ **to** $max(k, l)$ **do**
5   Compute $u^a_{i,s+j}$, $\mu^a_{i,s+j}$ and $\lambda^c_{i,s+j+1}$ using action ($\mathbb{W}^a$) and critic ($\mathbb{W}^c$) neural networks, respectively and $x_{i,s+j+1}$ by Eq. (10)
6   Compute λ^t_{ij}, u^t_{ij}, and μ^t_{ij} using Eqs. (12) and (14)
      $\mathbb{F}(u_{ij}, \mu_{ij}) = (\mathcal{L}_{u_{ij}}(z, \lambda, \sigma, \mu), -c(x_{ij}, u_{ij})) = 0$
7   **if** $j = M - 1$ **then**
8       $\mathbb{F}(u_{i,M-1}, \mu_{i,M-1}, \sigma_i) =$
      $(\mathcal{L}_{u_{i,M-1}}(z, \lambda, \sigma, \mu), -c(x_{i,M-1}, u_{i,M-1}), -\psi(x_{i,M}))$ with
      $\lambda_{iM} = \mathcal{G}_{x_{iM}}(x_{iM}) + \sigma_i \psi_{x_{iM}}(x_{iM})$
9      $err_c = \| \lambda^t_{ij} - \lambda^c_{ij} \|$
10      $err_a = \| (u, \mu)^t_{ij} - (u, \mu)^a_{ij} \|$
11   With the data set x_{ij}, λ^t_{ij} update the weight parameters $\mathbb{W}^c$
12   With the data set x_{ij}, $(u, \mu)^t_{ij}$ update the weight parameters $\mathbb{W}^a$
13   Set $\lambda^c_{ij} = \lambda^t_{i,j}$, $(u, \mu)^a_{i,j} = (u, \mu)^t_{i,j}$
14   Set $\hat{\lambda}_{i,j} = \lambda^t_{i,j}$, $(\hat{u}_{i,j}, \hat{\mu}_{i,j}) = (u, \mu)^t_{i,j}$
15   Compute $\hat{x}_{i,j+1}$ using Eq. (10) and $\hat{u}_{i,j}$
16    $\lambda_{0j} = \lambda_{1j}$, $\lambda_{Nj} = \lambda_{N-1,j}$
17   **return** $\hat{\lambda}_{i,j}$, $\hat{u}_{i,j}$, $\hat{\mu}_{i,j}$, $\hat{x}_{i,j+1}$

---

given by $z_k = tanh\left(\sum_{i=0}^{n} v_{ki}x_i\right), k = 1, \ldots, r$, where tanh(.) is the activation function for the hidden units. The response of the jth output unit is given by $y_j = \sum_{k=0}^{r} w_{jk}z_k, j = 1, \ldots, m$. The multiple layers of neurons with nonlinear transfer functions allow the network to learn nonlinear and linear relationships between the input and output vectors. The number of neurons in the input and output layers is given by the number of input and output variables, respectively. The multi-layered feed forward network is trained using the steepest descent error backpropagation rule. We can state the algorithm to solve the optimal control problem using the adaptive critic and recurrent neural network. In the Pontryagin's maximum principle for deriving an optimal control law, the interdependence of the state, costate and control dynamics is made clear. Indeed, the optimal control $\hat{u}$ and multiplier $\hat{\mu}$ is given by Eq. (14), while the costate Eqs. (12) and (13) evolve backward in time and depend on the state and control. The adaptive critic neural networks [12] are based on this relationship and consist of two networks at each node: an action network, the inputs for which are the current states and its outputs are the corresponding control $\hat{u}$ and multiplier $\hat{\mu}$, and the critic network for which the current states are inputs and current costates are outputs for normalizing the inputs and targets (zero mean and standard deviations). For detail explanation see [13]. The adaptive critic neural network procedure of the optimal control problem is summarized in Algorithm 1. In the adaptive critic synthesis, the action and critic network were selected such that they consist of $n+m$ subnetworks, respectively, each having $n-3n-1$ structure (i.e. n neurons in the input layer, $3n$ neurons in the hidden layer and one neuron in the output layer). The training procedure for the action and critic networks, respectively, are given by [12]. From the free terminal condition ($\psi(x) \equiv 0$) from Eqs. (12) and (13) we obtain that $\lambda_0 = -1$ and $\lambda_{iM} = 0, i = 1, \ldots, N$. We use this observation before proceeding to the actual training of the adaptive critic neural network. Further discussion and detail explanation of these adaptive critic methods can be found in [12,13].

5   Model of Feeding Adaptation of Daphnia

The model is described by a system of six delay partial differential equations. It is derived from the models of the series AQUAMOD [8,14] and modified by the inclusion of several algae species, diffusion and discrete time delay τ of food uptake by Daphnia. Four species of algae x_2, x_3, x_4, x_5 were considered during the computations performed. Each algae species are represented by a particular algal cell (or colony) volume. The volumes were set arbitrarily to ($V_i = 50, 500, 2500$ and $5000 \ \mu m^3$) to approximate the set of "edible" algal sizes commonly occurring in our reservoirs. The rates at which Daphnia and four algae species consume food are modelled by the Michaelis-Menten function with maximum rates $C_i, P_i, i = 2, \ldots, 5$, respectively, and half-maximum rates $a_4, s_i, i = 2, \ldots, 5$, respectively. The algae species respiration rates $r_i, i = 2, \ldots, 5$ are also include in the model. These ecological parameters of algae species are considered as functions of V_i. Functions occurring in the model are given in Table 1. The selectivity E_i is given by the following formula:

Table 1. Size-specific parameters of algae.

V_i	Algal cell volume $[\mu m^3)]$
$u_i = 2^{1/3}\sqrt[3]{\frac{3V_i}{4\pi}}$	Diameter corr. to V_i
$E_i(u) = exp(-0.1\,(u-u_i)^2)$	Selectivity
$C_i(u) = a_9 E_i(u)$	Forcing function
$P_i = 0.5 - 0.05 LOG(V_i)$	Spec. growth rate $[d^{-1}]$
$r_i = 0.02 + 0.002 LOG(V_i)$	Spec. resp. rate of algae $[d^{-1}]$
$s_i = -5 + 10 LOG(V_i)$	Half sat. constant for P $[mg.m^{-3}P]$
$i = 2,\ldots,5$	
$l_1(t) = 0.8 + 0.25cos(\frac{2\pi}{360}t)+$ $0.12cos(\frac{2\pi}{360}2t)$	Sedimentation function
$l_2(2) = 12 + 10sin(\frac{2\pi}{360}(t+220))$	Water temperature $[°C]$
$l_3(t) = 280 + 210sin(\frac{2\pi}{360}(t+240))$	Light intensity $[cal.cm^{-2}.day^{-1}]$
$l_4(t) = e^{(0.09 l_2(t))}$	
$l_5(t) = \frac{l_3(t)}{l_3(t)+a_{10}}$	
$d_1(t) = l_4(t)l_5(t)$	
$d_2(t) = a_1 l_1(t)$	
$d_3 = a_3 d_4$	
$d_4 = a_2 a_4$	

$$E_i(u) = exp(-0.1\,(u-u_i)^2), \tag{15}$$

where u is the value of setal density directly related to the algal diameter for which selectivity is maximal and $u_i = 2^{1/3}\sqrt[3]{\frac{3V_i}{4\pi}}$ is the diameter corresponding to each algal cell volume V_i. Because of the nonlinear relationship between diameters and algal cell volume ($V_i = 4/3\pi(u_i/2)^3$) the normal distribution given by Eq. (15) converts to a shape of a log-normal character. The specific filtration rate of algae of different sizes (volumes) of the population adapted to a certain condition (i.e. with certain values) of u becomes $C_i(u) = a_9 E_i(u)$, where $0 \leq a_9$ is the filtration rate for algae of optimal size, i.e. those which are filtered with the selectivity factors $E_i(u_i) = 1$.

The following system of partial functional differential equations with discrete time delay is proposed as a model of simple ecosystem:

$$\frac{\partial x_i(p,t)}{\partial t} = D_i \frac{\partial^2 x_i(p,t)}{\partial p^2} + F_i\,(x(p,t), x(p,t-\tau),t)\,,\; i = 1,\ldots,6, \tag{16}$$

where

$$F_1\,(x(p,t), x(p,t-\tau),t) = a_7(a_8 - x_1(p,t))$$
$$- d_1 x_1(p,t)\sum_{i=2}^{5}\frac{P_i x_i(p,t)}{x_1(p,t)+s_i} + l_2\sum_{i=2}^{5} r_i x_i(p,t)$$

$$+ x_6(p,t) \sum_{i=2}^{5} C_i x_i(p,t) \left(1 - \frac{d_4}{a_4 + x_i(p,t)}\right),$$

$$F_i\left(x(p,t), x(p, t-\tau), t\right) = \frac{d_1 P_i x_1(p,t) x_i(p,t)}{x_1(p,t) + s_i} - l_2 r_i x_i(p,t)$$
$$- E_i x_i(p,t) x_6(p,t) - d_2 x_i(p,t) + a_{i+9} a_7,$$
$$i = 2, \ldots, 5,$$

$$F_6\left(x(p,t), x(p, t-\tau), t\right) = d_3 x_6(p, t-\tau) \sum_{i=2}^{5} \frac{C_i x_i(p, t-\tau)}{a_4 + x_i(p, t-\tau)}$$
$$- a_5 x_6(p,t) + a_7 a_6,$$

with Neumann boundary condition

$$\frac{\partial x_i}{\partial p}(0,t) = \frac{\partial x_i}{\partial p}(1,t) = 0 \tag{17}$$

and initial conditions

$$x_i(p,t) = \phi_i(p,t) \geq 0, \ 0 \leq p \leq 1, \ t \in \langle -\tau, 0 \rangle, \ i = 1, \ldots, 6. \tag{18}$$

Here t denotes the time, p represents the spatial location, D_i are the diffusion coefficients and $x_i(t,p)$, i=1,...,6 are the concentration of phosphorus, four species of algae and zooplankton, respectively at time t and in spatial location p. The constant τ stands for the discrete time delay in uptake of algal species by Daphnia.

5.1   Optimization of Feeding Adaptation

In this section we are interested in the ability of Daphnia to adapt both the filtration area and filter density to the amount and size structure of the food particles of the (algae) population. We assume that the filtration in the aquatic filter feeders is an optimal process of maximal feeding strategy. We will investigate two strategies [8]:

(1) Instantaneous maximal biomass production as a goal function (local optimality), i.e.

$$\dot{x}_6 = F_6(x, u, t) \rightarrow max$$

for all t, under the constraints

$$u \in \langle u_{min}, u_{max} \rangle.$$

In the case of strategy (1), we maximize the following function

$$\mathbb{J}(u) = \sum_{i=2}^{5} \frac{a_9 E_i(u) d_3 x_i}{x_i + a_4}.$$

(2) Integral maximal biomass production as a goal function (global optimality), i.e.

$$\mathcal{J}(\hat{u}) = \int_a^b \int_0^{t_f} x_6(p,t) \, dt dp$$

under the constraints

$$u \in \langle u_{min}, u_{max} \rangle.$$

In the case of strategy (2), we have the following optimal control problem: to find a function $\hat{u}(p,t)$, for which the goal function $\mathcal{J}(u)$ attains its maximum. In the adaptive critic synthesis, the critic and action network were selected in a way that they consist of six and two subnetworks, respectively, each having 6-18-1 structure (i.e. six neurons in the input layer, eighteen neurons in the hidden layer and one neuron in the output layer).

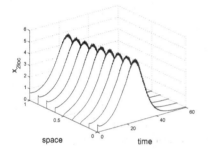 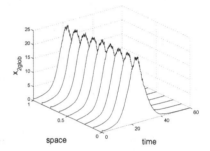

Fig. 1. Adaptive critic neural network simulation of optimal trajectory $\bar{x}_2(p,t)$ and $\hat{x}_2(p,t)$ with initial condition $\psi_s(t) = (0.1, 0.1, 0.2, 0.8, 0.4, 0.5, 0.6, 0.7, .1)\cos(2\pi p)$ for $t \in [-1, 0]$.

The proposed adaptive critic neural network is able to meet the convergence tolerance values we chose, leading to satisfactory simulation results. The MATLAB simulations show that the proposed neural network is able to solve the nonlinear optimal control problem with state and control constraints. Our results are quite similar to those obtained in [8]. The results of numerical solutions (Figs. 1 and 2) have shown that the optimal trajectories $\bar{x}(p,t)$ and $\hat{x}(p,t)$ based on a short and long-term perspective, respectively are different. When $\hat{u}(p,t)$ is optimal then $\mathcal{J}(\hat{u}(p,t)) \geq \mathbb{J}(\tilde{u}(p,t))$. The numerical results show that for the initial conditions considered the total biomass for the short-term perspective is smaller than the biomass for the long-term perspective, i.e. $\mathcal{J}(\hat{u}(p,t)) > \mathbb{J}(\tilde{u}(p,t))$. The higher biomass of zooplankton is obtained in case of integral formulation points towards the assumption that the organisms do better if not reacting only to the immediate changes, but having developed mechanisms consistent with more long-term consideration. The numerical solutions show that in the case of optimal strategies $\hat{u}(p,t))$, $\tilde{u}(p,t)$ we have different time trajectories and all species of algae are able to survive.

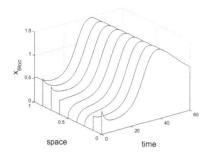 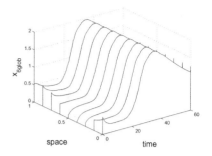

Fig. 2. Adaptive critic neural network simulation of optimal trajectory $\bar{x}_6(p,t)$ and $\hat{x}_6(p,t)$ with initial condition $\psi_s(t) = (0.1, 0.1, 0.2, 0.8, 0.4, 0.5, 0.6, 0.7, .1)\cos(2\pi p)$ for $t \in [-1, 0]$.

6   Conclusion

In this paper, the algorithm of neural networks based solution of parabolic distributed optimal control problems was presented. To better illustrate these advances in computational optimization, we gave a detailed description of discretization of distributed optimal control problems by finite differences and solved by means of time-space multigrid techniques, where co-state variable was approximated by feed forward neural networks. To illustrate the algorithm we considered a feeding adaptation of filter feeders of Daphnia. We formulated, analysed and solved a distributed optimal control problem related to the optimal uptake of nutrient by Daphnia. The numerical simulations by using MATLAB have shown that the adaptive critic neural network is able to solve the nonlinear optimal control problem with control and state constraints. Numerical simulations have also shown that in the case of local and global optimal strategies, respectively, we have different time trajectories and varying filter density as a result of a local and global optimality, which explains the adaptation of filter feeders of Daphnia.

Acknowledgments. The paper was worked out as a part of the solution of the scientific project number KEGA 010UJS-4/2014.

References

1. Borzi, A.: Multigrid methods for parabolic distributed optimal control problems. J. Comput. Appl. Math. **157**, 365–382 (2003)
2. Buskens, C., Maurer, H.: SQP-methods for solving optimal control problems with control and state constraints: adjoint variable, sensitivity analysis and real-time control. J. Comput. Appl. Math. **120**, 85–108 (2000)
3. Chryssoverghi, I.: Discretization methods for semilinear parabolic optimal control problems. Int. J. Numer. Anal. Model. **3**, 437–458 (2006)

4. Clever, D., Lang, J., Ulbrich, S., Ziems, J.C.: Combination of an adaptive multilevel SQP method and a space-time adaptive PDAE solver for optimal control problems. Procedia Comput. Sci. **1**, 1435–1443 (2012)
5. Gollman, L., Kern, D., Mauer, H.: Optimal control problem with delays in state and control variables subject to mixed control-state constraints. Optim. Control Appl. Meth. **30**, 341–365 (2009)
6. Hornik, M., Stichcombe, M., White, H.: Multilayer feed forward networks are universal approximators. Neural Netw. **3**, 256–366 (1989)
7. Kmet, T., Kmetova, M.: Adaptive critic neural network solution of optimal control problems with discrete time delays. In: Mladenov, V., Koprinkova-Hristova, P., Palm, G., Villa, A.E.P., Appollini, B., Kasabov, N. (eds.) ICANN 2013. LNCS, vol. 8131, pp. 483–494. Springer, Heidelberg (2013)
8. Kmet, T., Straskraba, M.: Feeding adaptation of filter feeders: Daphnia. Ecol. Model. **178**, 313–327 (2004)
9. Knowles, G.: Finite element approximation of parabolic time optimal control problems. SIAM J. Control Optim. **20**, 414–427 (1982)
10. Liu, W., Yan, N.: A posteriori error estimates for optimal control problems governed by parabolic equations. Numer. Math. **93**, 497–521 (2003)
11. Mittelmann, H.D.: Solving elliptic control problems with interior point and SQP methods: control and state constraints. J. Comput. Appl. Math. **120**, 175–195 (2000)
12. Padhi, R., Unnikrishnan, N., Wang, X., Balakrishnan, S.N.: Adaptive-critic based optimal control synthesis for distributed parameter systems. Automatica **37**, 1223–1234 (2001)
13. Rumelhart, D.F., Hinton, G.E., Wiliams, R.J.: Learning internal representation by error propagation. In: Rumelhart, D.E., McClelland, D.E., Group, P.R. (eds.) Parallel Distributed Processing: Foundation Observation of Strains, vol. 1, pp. 318–362. The MIT Press, Cambridge (1987)
14. Straskraba, M., Gnauck, P.: Freshwater Ecosystems: Developments in Environmental Modelling, Modelling and Simulation. Elsevier, Amsterdam (1985)
15. Werbos, P.J.: Approximate dynamic programming for real-time control and neural modelling. In: White, D.A., Sofge, D.A. (eds.) Handbook of Intelligent Control: Neural Fuzzy, and Adaptive Approaches, pp. 493–525. Van Nostrand, New York (1992)

Generating the Logicome of a Biological Network

Charmi Panchal, Sepinoud Azimi, and Ion Petre[(⊠)]

Computational Biomodeling Laboratory, Turku Centre for Computer Science,
Åbo Akademi University, Agora, Domkyrkotorget 3, 20500 Åbo, Finland
{cpanchal,sazimi,ipetre}@abo.fi

Abstract. There has been much progress in recent years towards building larger and larger computational models for biochemical networks, driven by advances both in high throughput data techniques, and in computational modeling and simulation. Such models are often given as unstructured lists of species and interactions between them, making it very difficult to understand the *logicome* of the network, i.e. the logical connections describing the activation of its key nodes. The problem we are addressing here is to predict whether these key nodes will get activated at any point during a fixed time interval (even transiently), depending on their initial activation status. We solve the problem in terms of a Boolean network over the key nodes, that we call the logicome of the biochemical network. The main advantage of the logicome is that it allows the modeler to focus on a well-chosen small set of key nodes, while abstracting away from the rest of the model, seen as biochemical implementation details of the model. We validate our results by showing that the interpretation of the obtained logicome is in line with literature-based knowledge of the EGFR signalling pathway.

Keywords: Biomodeling · Boolean network · Logicome · EGFR pathway · ODE models

1 Introduction

One of the central topics of interest in systems biology is to identify the functionalities of a living cell and to understand how the huge number of interactions within a cell facilitate such functionalities. The set of complex and involved interactions lead to obtaining a large number of collected experimental data as well as complex networks. These broad sources of information can prove to be very useful in providing a realistic life picture of the phenomenon under study, but can also make it difficult to analyze the system and can cause inaccuracy in predicting the system's behavior. Identifying the main players within a network and understanding how they activate each other can help to overcome these difficulties.

There have been many studies on the logical modelling of biological networks; for example, [4–6,30] discuss the correspondence between Boolean networks and ODEs; for an introduction to Boolean networks and ODEs we refer

© Springer International Publishing Switzerland 2016
M. Botón-Fernández et al. (Eds.): AlCoB 2016, LNBI 9702, pp. 38–49, 2016.
DOI: 10.1007/978-3-319-38827-4_4

to [13,14] respectively. Fuzzy logic was used in [19] to yield the logical models corresponding to the biological networks. As a different approach, [27] build the Boolean logic models by training a literature-based prior knowledge network against biochemical data. These studies mainly proposed approaches where the full understanding of the biological aspects of the phenomenon under study was crucial and the goal was to obtain a mathematical model reproducing that understanding. Our study goes in the reverse direction: it starts from an existing mathematical model and aims to obtain an abstract, high-level understanding of the functionality of the biological network underlying the model. Our goal is to obtain a logical description of the activation conditions between the key nodes of the network; even in the case when one starts from a detailed biological model going towards the mathematical model, our reverse engineering approach brings a new higher-level understanding of the functionality of the biological model we started from. The result of our approach is formulated as a Boolean network whose nodes are the key species we focus on; we coin the term *logicome* to name this network.

Extracting a Boolean network model from a given ODE-based model is a well-studied topic with many different solutions, see, e.g., [30] for a recent new solution and a good overview of the topic. Typically, the Boolean network model is seen as a companion of the ODE-based model, compensating for the lack of detailed kinetic-level data for the model, or allowing for alternative global analysis of model dynamics, such as attractor- or multi-stability- analysis, see [30]. A key step going from an ODE model to its corresponding Boolean network model is the discretization scheme allowing to replace continuous variables with their corresponding 0/1 variables. This is typically done by sampling the numerical integration of the continuous variables at different time points and by discretizing their values at those points. This leads to the dynamics of the Boolean model being interpreted in terms of discrete time series reflecting the behavior of the original ODE model. Our approach is coarser: we aim to capture the activation of the key nodes of the model over the whole time interval (to be thought of as much larger than those involved in the discretization of ODE models). This includes capturing the transient activation of a node over that interval, even if at the extremities of the interval the node may be inactive. The result is a Boolean network that accompanies the starting ODE model in terms of describing asynchronous cause-effect relationships among its key nodes over a fixed time interval.

As a case study we focus on the *EGFR (epidermal growth factor receptor) signaling pathway*. Epidermal growth factors are key players in cell proliferation, survival, migration and differentiation. EGFR signaling also has a major role in EGFR-dependent signal transduction, see [29]. Therefore, understanding their behavior is crucial in any cancer related studies, see [20]. For more information on EGFR signaling pathways we refer to [2, 29, 32].

This paper is organized as follows. In Sect. 2, we present our methodology to infer the logicome of biochemical networks. In Sect. 3, we introduce the case study we used in this paper. In Sect. 4 we present the results of applying the method to the case study and analyze the produced results and finally we conclude with some discussions in Sect. 5. All the models and data files used in this paper can be found at: http://combio.abo.fi/research/logicome-models-2/.

2   Methodology

In this section we present our method to infer the logicome of an ODE-based model. The steps are described in a generic way – their detailed implementation is up to the modeler and it depends on the case study. In the next section we discuss one particular way in which we used this method in the case of the EGFR pathway.

Step 1 – Setup. We start with an ODE model for a biochemical network. We assume also to have a set of "key nodes" whose influences over each others' activation we aim to capture. The choice of the key nodes from among the variables of the ODE model depends on the modeler and on the network under study.

Step 2 – Discretization. To be able to describe the logicome of a network in terms of Boolean network, we need to translate continuous simulation data to a Boolean, "on/off"-based language. Therefore, as the second step we incorporate a discretization algorithm into our method. Many discretization methods exist, see for example [18, 26]. In this study our discretization step is based on a threshold-based approach in which we assign "1" to a species if at any time during the simulation its value is above a given threshold, and "0" otherwise. The precise choice of the threshold depends on the network under study.

Step 3 – Simulation. We simulate all possible knock-out mutants; in other words, all models where the key species are turned on/off in all possible combinations. We then apply to each simulation result the discretization step to obtain the Boolean results corresponding to each mutant. In this way we produce a truth table describing the output of each simulation as a Boolean function with the key nodes as its Boolean variables. Translating the input Boolean values of the key nodes to absolute numerical values to be used in the simulation can be done in several different ways, depending on the case study. For example, the 0 value for a Boolean key node may be translated to value 0 for the corresponding variable(s) in the knock-out mutant, while value 1 may be translated to the threshold value chosen for that variable in Step 2. The other, non-key nodes get the same initial values as in the original model.

Step 4 – Logicome generation. In this step we generate the logicome corresponding to the given biochemical network from the produced truth table in the previous step. Different algorithms can be used to implement this step, see for example [1, 11, 16, 21]. In this paper we use the *Logic Friday* tool which incorporates the *Espresso algorithm* proposed in [21].

3   Case-Study: The EGFR Pathway

We focus in this paper on a signaling network that is strongly associated with the development of cancer processes: the *EGFR signaling pathway*. In the following subsections we provide a brief biological background and some computational details of this model.

3.1   Biological Background

The epidermal growth factor receptor (EGFR) pathway regulates several important cellular processes including cell proliferation, survival, differentiation and development, see [20]. Because of its association with the various types of cancer processes, this pathway is a widely investigated signal transduction system. The EGFR pathway can be seen as a union of several smaller pathways, also called *modules*, see [3,31]. The proteins situated at the intersection between these modules are called *interface species*. The analysis presented in [10] identifies the locations of oncogenes and essential components of the EGFR signaling cascade that define most of the interface regions. Our model is adopted from [31] that uses the model originally presented in [28] and implements it in the stochastic pi-calculus language together with the results identified by [10]. We follow the approach of [31] and their modularization of the EGFR signaling pathway in the following 7 modules: EGF, Grb2, Ras-Shc-Dependent /Independent, Raf, MEK, and ERK. These modules communicate with each other through the following 8 *interface species*: (EGF-EGFR*)2-GAP, (EGF-EGFR*)2-GAP-Grb2-Sos, (EGF-EGFR*)2-GAP-Shc*-Grb2-Sos, Ras-GTP, Ras-GTP*, MEK-PP, Raf* and ERK-PP. We adopt these interface species as the key nodes in our approach.

We briefly describe the functionality of the EGFR pathway focusing mainly on the signal propagation within the interface species, as suggested in [10]; the modules of the pathway are considered as black-boxes communicating to each other through the interface species. The EGFR is situated on the extracellular surface of the cell and signal transduction begins upon binding of ligand EGF (epidermal growth factor) to EGFR. The EGF-bounded receptor induces dimerization and autophosphorylation of several members of intracellular domains, which leads to the recruiting of several cytoplasmic enzymes and adaptor proteins. This initiates to the activation of two principal pathways, one Shc-dependent and another Shc-independent, that play a significant role in the activation of downstream signaling processes like hydrolyzation of Ras-GDP and activation of Ras-GTP that follows by dissociation of Ras-GTP from the receptor complex. Further dissociation of Ras-GTP makes it inactive and promotes the intrinsic activity of Ras protein regulated by the GTPase activating protein (GAP) that is involved in several crucial cellular processes see [10,24]. It is assumed that the dissociated Ras-GTP molecule causes phosphorylation of the Raf protein that in-turn double phosphorylates MEK (turning it to MEK-PP) and ERK (turning it to ERK-PP) proteins. The final result of the signaling cascade is the double phosphorylated ERK-PP that further regulates a number of transcription factors and essential proteins for cell differentiation and growth.

A systematic analysis of control mechanisms (including positive/negative feedback loops) underlying EGFR pathway are presented in [10,31]. We aim to represent the functional relationships associated with the interface species through a Boolean network – the *logicome* of the EGFR signaling pathway.

3.2   Mathematical Model, Simulation and Discretization

We associated a mass-action ODE-based model, see [8,14], to the reaction based model of [10]. Each of the 103 variable molecular species of the model in [10] gets a variable in our mathematical model. We wrote the reaction-based model of the EGFR pathway in the COPASI software, see [9], and used its feature to automatically generate the mass-action-based system of ordinary differential equations associated to the model. We call the resulting model our *basic model*.

Following the approach of [31], we simulated in COPASI this model for an EGF stimulus of 4981 molecules/pl which is enough to phosphorylate 50000 EGF-receptors. The simulation was run for 6000 s and the time series results of each interface species were collected.

For our method we are interested in analyzing all knock-out mutants where the interface species are active/inactive in all possible combinations. In the knock-out mutants the initial values of the inactive interface species are set to the value 0, while the active interface species are set to a specific threshold value of 1 % of that species' maximum value in the simulation of the basic model up to 6000 s. Since we considered 8 interface species, we have $256 = 2^8$ knock-out mutant simulations.

3.3   Generating the Logicome

Each knock-out mutant can be seen as a particular truth assignment over the 8 Boolean variables standing for the interface species. The results of the 256 knock-out simulations were discretized as follows.

Collecting the outputs of all knock-out mutants can be done in the form of a Boolean function with 8 inputs and 8 outputs.

We used the *LogicFriday* software to generate the Boolean function associated to the EGFR pathway based on the Boolean table collected above. We then used the 5 types of Boolean gates illustrated in Fig. 1 to generate the logicome associated to the EGFR signaling pathway.

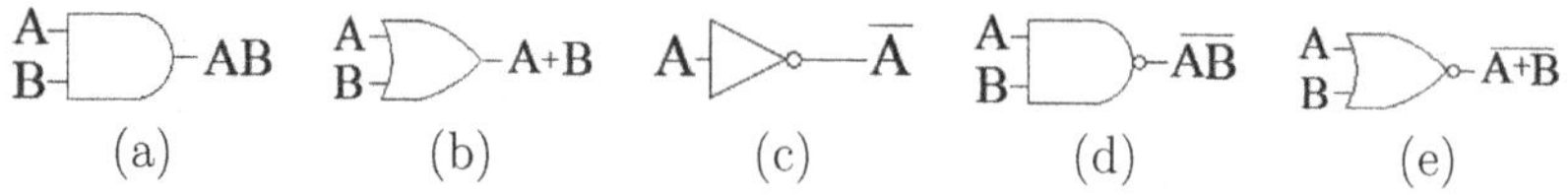

Fig. 1. The Boolean gates for the logical outcome: (a) AND : AB, (b) OR : $A + B$, (c) NOT : $\overline{A}$, (d) NAND: $\overline{AB}$, (e) NOR : $\overline{A + B}$, where we denote the negation of A with $\overline{A}$, the disjunction of A and B with $A + B$, and the conjunction of A and B with AB.

4   Results

The interface species are denoted in the logicome as the nodes of the Boolean network in the way explained in Table 1. The Boolean functions generated as the result of the steps described in Sect. 3 are shown in Table 2. We repeated the same experiment where we set the initial values of the active key nodes to 10 % (rather than 1 %) of their maximum value in the simulation of the basic model; the corresponding Boolean formulation is presented in Table 3.

Table 1. The notation used for the interface species in the Boolean network.

Node	Interface species
G_0	(EGF-EGFR*)2-GAP
G_1	Raf*
G_2	MEK-PP
G_3	Ras-GTP*
G_4	ERK-PP
G_5	(EGF-EGFR*)2-GAP-Shc*-Grb2-Sos
G_6	Ras-GTP
G_7	(EGF-EGFR*)2-GAP-Grb2-Sos

Table 2 shows G_1 as getting activated in all knock-out models and thus, being set to constant 1. This means that for all combinations of active/inactive key nodes (even those where G_1 is initialized as inactive), G_1 gets eventually activated in the time interval $[0, 6000]$ sec. This can be interpreted as G_1 being insensitive to (relatively) small changes in the levels of the other key nodes; indeed, all the key nodes are 0 in the basic model, leading to activation of G_1; setting the initial values of the key nodes to 1 % of their maximum level in the basic model does not change the situation. This result also suggests that in the case of small perturbations in the initial values of key nodes, the activation of G_1 is driven by other factors, outside the set of key nodes. The situation is different if we look into bigger changes in the initial values of the key nodes, e.g., setting them to 10 % of their maximum values in the basic model; as shown in Table 3, G_1 is in this case non-constant and influencing the behavior of G_6. In Table 3, we observe that the activation of G_1 depends on the key nodes G_3, G_5 and G_6 – this is consistent with the results reported in [25].

Another interesting observation of the logicome in Table 2 is that all key nodes get activated in the case of G_3 starts inactive and G_5 starts active. The same observation is found in the results obtained for the threshold of 10 %, see Table 3, and even for 20 % and 30 % see Tables 4 and 5. This is consistent with the observation of [7,10,23,31] about the role played by the shc*-dependent component (denoted by G_5) and the Ras subfamily protein (denoted by G_3) in the activation of several pathway components, including all of our key nodes.

Table 2. The Boolean functions describing the logicome of the EGFR signaling pathway for the threshold of 1 %. An overline over a variable's name denotes its negation, the plus denotes disjunction, while the concatenation of two variables denotes their conjunction.

Boolean functions
$G_0 := \overline{G}_3 + G_5 + G_0\overline{G}_4 + \overline{G}_4 G_7 + G_0\overline{G}_6 G_7;$
$G_1 := 1;$
$G_2 := G_2 + \overline{G}_3 + G_5 + G_6;$
$G_3 := G_0 + \overline{G}_2 + G_3 + G_4 + G_5 + G_6 + G_7;$
$G_4 := G_2 + \overline{G}_3 + G_4 + G_6 + G_0 G_5 G_7;$
$G_5 := G_0 G_5 + \overline{G}_3 G_5 + \overline{G}_3\overline{G}_6 + G_5\overline{G}_6 + G_5 G_7 + G_0\overline{G}_3 G_7;$
$G_6 := \overline{G}_3 + G_5 + G_0\overline{G}_6 + G_6 G_7;$
$G_7 := \overline{G}_3 + G_5$

Table 3. The Boolean functions describing the logicome of the EGFR signaling pathway for the threshold of 10 %.

Boolean functions
$G_0 := G_5 + G_0\overline{G}_3\overline{G}_4 + \overline{G}_3\overline{G}_4\overline{G}_6 + G_0\overline{G}_3 G_7 + \overline{G}_3\overline{G}_4 G_7;$
$G_1 := \overline{G}_3 + G_5 + G_6;$
$G_2 := G_2 + \overline{G}_3 + G_5 + G_6;$
$G_3 := G_0 + \overline{G}_2 + G_3 + G_5 + G_6 + G_7;$
$G_4 := G_2 + \overline{G}_3 + G_4 + G_6 + G_0 G_5 G_7;$
$G_5 := G_0 G_5 + \overline{G}_3 G_5 + \overline{G}_3\overline{G}_6 + G_5\overline{G}_6 + G_5 G_7 + G_0\overline{G}_3 G_7;$
$G_6 := G_5 + G_0\overline{G}_3 + \overline{G}_1\overline{G}_3 + G_0 G_6 + \overline{G}_3 G_6 + \overline{G}_3 G_7 + G_6 G_7;$
$G_7 := \overline{G}_3 G_5 + \overline{G}_3\overline{G}_6 + \overline{G}_3 G_7 + G_0 G_5\overline{G}_6 + G_0 G_5 G_7 + G_5\overline{G}_6 G_7$

It is also interesting to note that the EGFR signaling pathway has an internal mechanism for compensating the potential failure of G_5 by G_7. Based on [7,10, 31], G_0 mediates the activation of both G_5 and G_7; in case G_5 fails while G_3 remains inactive then G_7 gets activated and this is enough to activate all key nodes. This is seen in Table 3, if $G_0 = \overline{G}_3 = \overline{G}_5 = G_7 = 1$, then all key nodes get activated.

4.1 Sensitivity to the Numerical Setup of the Model

To investigate the sensitivity of our method to changes in the numerical setups of the underlying ODE model, we re-ran all simulations for different values of EGF and EGFR. We first experimented with different concentrations of EGF stimulus keeping the same EGFR concentration of 50000 molecules and then with different concentrations of EGFR keeping the same EGF stimulus of 4981 molecules. We observe that the obtained logicomes are almost identical to the previous result

Table 4. The Boolean functions describing the logicome of the EGFR signaling pathway for the threshold of 20 %.

Boolean functions
$G_0 := G_5 + G_0\overline{G}_3\overline{G}_4 + \overline{G}_3\overline{G}_4\overline{G}_6 + G_0\overline{G}_3G_7 + \overline{G}_3\overline{G}_4G_7;$
$G1 := \overline{G}_3 + G_5 + G_6;$
$G_2 := G_2 + \overline{G}_3 + G_5 + G_6;$
$G_3 := G_0 + \overline{G}_2 + G_3 + G_5 + G_6 + G_7;$
$G_4 := G_2 + \overline{G}_3 + G_4 + G_6 + G_0G_5G_7;$
$G_5 := G_0G_5 + \overline{G}_3G_5 + \overline{G}_3\overline{G}_6 + G_5\overline{G}_6 + G_5G_7 + G_0\overline{G}_3G_7;$
$G_6 := G_5 + G_0\overline{G}_3 + \overline{G}_1\overline{G}_3 + G_0G_6 + \overline{G}_3G_6 + \overline{G}_3G_7 + G_6G_7;$
$G_7 := \overline{G}_3G_5 + \overline{G}_3\overline{G}_6 + \overline{G}_3G_7 + G_0G_5G_7 + G_5\overline{G}_6G_7$

Table 5. The Boolean functions describing the logicome of the EGFR signaling pathway for the threshold of 30 %.

Boolean functions
$G_0 := G_5 + G_0\overline{G}_3\overline{G}_4 + \overline{G}_3\overline{G}_4\overline{G}_6 + G_0\overline{G}_3G_7 + \overline{G}_3\overline{G}_4G_7;$
$G_1 := \overline{G}_3 + G_5 + G_6;$
$G_2 := G_2 + \overline{G}_3 + G_5 + G_6;$
$G_3 := G_0 + G_3 + G_5 + G_6 + G_7 + \overline{G}_1\overline{G}_2 + \overline{G}_2G_4;$
$G_4 := G_2 + \overline{G}_3 + G_4 + G_6 + G_0G_5G_7;$
$G_5 := G_0G_5 + \overline{G}_3G_5 + \overline{G}_3\overline{G}_6 + G_5\overline{G}_6 + G_5G_7 + G_0\overline{G}_3G_7;$
$G_6 := G_5 + G_0\overline{G}_3 + \overline{G}_1\overline{G}_3 + G_0G_6 + \overline{G}_3G_6 + \overline{G}_3G_7 + G_6G_7;$
$G_7 := \overline{G}_3G_5 + \overline{G}_3\overline{G}_6 + \overline{G}_3G_7 + G_0G_5G_7 + G_5\overline{G}_6G_7$

presented in Table 2. To investigate the sensitivity of our method to different threshold criteria, we repeated the experiments above with a threshold value of 30 % of each interface species' maximum value. By comparing results, we note that the logicome results obtained with the threshold value of 10 %, 20 %, and 30 % (see Tables 3, 4, and 5) are much more complex than the previous one.

4.2   Incomplete Availability of the Knock-Out Mutants

In the way we described our method in Sects. 2 and 3, we implicitly assume the full availability of the simulation results of all knock-out mutant models. We considered the case when the data on several knock-out mutants is in fact not available and compared the results to the case when all data is available. We considered the simulations results of only 186 knock-out mutants and assumed that the data on the other 70 knock-out mutants is unavailable. We used the threshold value of 1 % and the numerical setups of EGF and EGFR as 4981 and 50000 molecules, respectively.

Table 6. The Boolean functions associated with the logicome of the model where the data of 70 knockout mutants are not available. The result is almost identical to that in Table 2 where all data was available, showing that the method in this case was robust to missing data.

Boolean functions
$G_0 := \overline{G}_3 + G_5 + G_0\overline{G}_4 + \overline{G}_4 G_7 + G_0\overline{G}_6 G_7;$
$G_1 := 1;$
$G_2 := G_2 + \overline{G}_3 + G_5 + G_6;$
$G_3 := \overline{G}_2 + G_3 + G_4 + G_5 + G_6 + G_7;$
$G_4 := G_2 + \overline{G}_3 + G_4 + G_6 + G_0 G_5 G_7;$
$G_5 := G_0 G_5 + \overline{G}_3 G_5 + \overline{G}_3\overline{G}_6 + G_5\overline{G}_6 + G_5 G_7 + G_0\overline{G}_3 G_7;$
$G_6 := \overline{G}_3 + G_5 + G_0 G_6 + G_6 G_7;$
$G_7 := \overline{G}_3 + G_5$

The result obtained in this case is shown in the Table 6 and it is almost the same as the result in Table 2 obtained by using the full data. This shows that in this case the logicome extraction method was robust to the missing data; this may of course be different for other models and for other missing data.

5 Discussion

We propose in this article an addition to the rich field of logic modeling of biological networks, see, e.g., [4,15,19]. We start from a mathematical model of the network, taking advantage of the growing availability of mathematical models. The logicome approach proposed in this article allows the modeler to focus on a selected set of key nodes, important for the network under study, while abstracting away from the rest of the network; the output is a description of their influence on each other (even transient) activation over a fixed time interval.

The bottom-up modeling approaches (e.g., large-scale modeling [17], automatic knowledge extraction [22], data-driven network construction [12], etc.) have been very popular due to their ability to provide a very detailed picture, to explain the data, and to reproduce the behaviour of the phenomenon under study. The logicome is a companion to such detailed models; it gives a more abstract, systematic and objective description of the functionalities of the model. This is especially relevant in the case of big models built from many different sub-models and for which a full global "blueprint" does not exist. The logicome aims to be such a blueprint, deduced a-posteriori, based on an existing detailed view of the model.

The output of the logicome approach depends on the numerical setup of the method: both on the numerical setup of the basic mathematical model, and on the choice of the threshold values in the discretization step. This is natural since the method is dependent on the numerical ODE-based simulations of the basic model and of the knock-out mutants; this suggests choosing an already well-fitted and -validated model for the network under study. The choice of the threshold value is in fact a decision on how a species of the model can be labeled as 'active'; we suggested using a percentage of the maximum value reached by that species in the simulation of the basic model, but other choices may also be appropriate depending on the case study.

The computational efficiency of the method is dependent on the number of key nodes selected in the analysis: with more key nodes selected, exponentially more knock-out mutant models should be analyzed. Eliminating some of the knock-out mutants is possible, and the result of the method will be in this case an only-partial description of the logical dependencies between the key nodes. On the other hand, the method scales up very well in the size of the basic model: as long as the ODE-based models may be simulated efficiently, the method will be practical; this means that networks with thousands of nodes may be analyzed, as long as the number of key nodes n is so that it remains practical to run 2^n simulations.

References

1. Akutsu, T., Miyano, S., Kuhara, S.: Identification of genetic networks from a small number of gene expression patterns under the Boolean network model. In: Somogyi, R., Kitano, H. (eds.) Pacific Symposium on Biocomputing, vol. 4, pp. 17–28. Citeseer (1999)
2. Britton, D., Hutcheson, I.R., Knowlden, J.M., Barrow, D., Giles, M., McClelland, R.A., Gee, J.M., Nicholson, R.I.: Bidirectional cross talk between ERα and EGFR signalling pathways regulates tamoxifen-resistant growth. Breast Cancer Res. Treat. **96**(2), 131–146 (2006)
3. Bruggeman, F.J., Westerhoff, H.V., Hoek, J.B., Kholodenko, B.N.: Modular response analysis of cellular regulatory networks. J. Theor. Biol. **218**(4), 507–520 (2002)
4. Chaves, M., Sontag, E.D., Albert, R.: Methods of robustness analysis for Boolean models of gene control networks. IEEE Proc. Syst. Biol. **153**(4), 154–167 (2006)
5. Davidich, M.I., Bornholdt, S.: Boolean network model predicts cell cycle sequence of fission yeast. PloS ONE **3**(2), e1672 (2008)
6. Glass, L., Kauffman, S.A.: The logical analysis of continuous, non-linear biochemical control networks. J. Theor. Biol. **39**(1), 103–129 (1973)
7. Gong, Y., Zhao, X.: Shc-dependent pathway is redundant but dominant in mapk cascade activation by egf receptors: a modeling inference. FEBS Lett. **554**(3), 467–472 (2003)
8. Gratie, D.-E., Iancu, B., Petre, I.: ODE analysis of biological systems. In: Bernardo, M., de Vink, E., Di Pierro, A., Wiklicky, H. (eds.) SFM 2013. LNCS, vol. 7938, pp. 29–62. Springer, Heidelberg (2013)
9. Hoops, S., Sahle, S., Gauges, R., Lee, C., Pahle, J., Simus, N., Singhal, M., Xu, L., Mendes, P., Kummer, U.: COPASI - a complex pathway simulator. Bioinformatics **22**(24), 3067–3074 (2006)

10. Hornberg, J.J., Binder, B., Bruggeman, F.J., Schoeberl, B., Heinrich, R., Westerhoff, H.V.: Control of MAPK signalling: from complexity to what really matters. Oncogene **24**(36), 5533–5542 (2005)
11. Hwa, H.R.: A method for generating prime implicants of a Boolean expression. IEEE Trans. Comput. **23**(6), 637–641 (1974)
12. Janes, K.A., Yaffe, M.B.: Data-driven modelling of signal-transduction networks. Nat. Rev. Mol. Cell Biol. **7**(11), 820–828 (2006)
13. Kauffman, S.: Homeostasis and differentiation in random genetic control networks. Nature **224**, 177–178 (1969)
14. Klipp, E., Herwig, R., Kowald, A., Wierling, C., Lehrach, H.: Systems Biology in Practice: Concepts, Implementation and Application. Wiley, Weinheim (2008)
15. Le Novere, N.: Quantitative and logic modelling of molecular and gene networks. Nat. Rev. Genet. **16**(3), 146–158 (2015)
16. Liang, S., Fuhrman, S., Somogyi, R.: Reveal, a general reverse engineering algorithm for inference of genetic network architectures. In: Bryant, B., Milosavljevic, A., Somogyi, R. (eds.)Pacific Symposium on Biocomputing, vol. 3, pp. 18–29. Citeseer (1998)
17. Macklin, D.N., Ruggero, N.A., Covert, M.W.: The future of whole-cell modeling. Curr. Opin. Biotechnol. **28**, 111–115 (2014)
18. Martin, S., Zhang, Z., Martino, A., Faulon, J.: Boolean dynamics of genetic regulatory networks inferred from microarray time series data. Bioinformatics **23**(7), 866–874 (2007)
19. Morris, M.K., Saez-Rodriguez, J., Sorger, P.K., Lauffenburger, D.A.: Logic-based models for the analysis of cell signalling networks. Biochemistry **49**(15), 3216–3224 (2010)
20. Oda, K., Matsuoka, Y., Funahashi, A., Kitano, H.: A comprehensive pathway map of epidermal growth factor receptor signaling. Curr. Opin. Biotechnol. **1**(1), 1–17 (2005)
21. Pantel, P., Pennacchiotti, M.: Espresso: Leveraging generic patterns for automatically harvesting semantic relations. In: Carpuat, M., Duh, K. (eds.) Proceedings of the 21st International Conference on Computational Linguistics and the 44th Annual Meeting of the Association for Computational Linguistics, Association for Computational Linguistics, pp. 113–120 (2006)
22. Pitkänen, E., Jouhten, P., Hou, J., Syed, M.F., Blomberg, P., Kludas, J., Oja, M., Holm, L., Penttilä, M., Rousu, J., Arvas, M.: Comparative genome-scale reconstruction of gapless metabolic networks for present and ancestral species. PLoS Comput. Biol. **10**(2), 1–12 (2014)
23. Rajalingam, K., Schreck, R., Rapp, U.R., Albert, V.: Ras oncogenes and their downstream targets. Biochimica et Biophysica Acta (BBA) - Molecular Cell Research **1773**(8), 1177–1195 (2007)
24. Rajasekharan, S., Raman, T.: Ras and ras mutations in cancer. Cent. Eur. J. Biol. **8**(7), 609–624 (2013)
25. Roskoski, R.: Raf protein-serine/threonine kinases: structure and regulation. Biochem. Biophys. Res. Commun. **399**(3), 313–317 (2010)
26. Saez-Rodriguez, J., Alexopoulos, L.G., Epperlein, J., Samaga, R., Lauffenburger, D.A., Klamt, S., Sorger, P.K.: Discrete logic modelling as a means to link protein signalling networks with functional analysis of mammalian signal transduction. Mol. Syst. Biol. **5**(1), 331 (2009)
27. Saez-Rodriguez, J., Alexopoulos, L.G., Zhang, M., Morris, M.K., Lauffenburger, D.A., Sorger, P.K.: Comparing signaling networks between normal and transformed hepatocytes using discrete logical models. Cancer Res. **71**(16), 5400–5411 (2011)

28. Schoeberl, B., Eichler-Jonsson, C., Gilles, E.D., Müller, G.: Computational modeling of the dynamics of the MAP kinase cascade activated by surface and internalized EGF receptors. Nat. Biotechnol. **20**(118), 370–375 (2002)

29. Sebastian, S., Settleman, J., Reshkin, S.J., Azzariti, A., Bellizzi, A., Paradiso, A.: The complexity of targeting EGFR signalling in cancer: from expression to turnover. Biochimica et Biophysica Acta (BBA)-Reviews on Cancer **1766**(1), 120–139 (2006)

30. Stötzel, C., Röblitz, S., Siebert, H.: Complementing ODE-based system analysis using Boolean networks derived from an Euler-like transformation. PLoS ONE **10**(10), e0140954 (2015)

31. Wang, D.Y., Cardelli, L., Phillips, A., Piterman, N., Fisher, J.: Computational modeling of the EGFR network elucidates control mechanisms regulating signal dynamics. BMC Syst. Biol. **3**(1), 1–18 (2009)

32. Yarden, Y.: The EGFR family and its ligands in human cancer: signalling mechanisms and therapeutic opportunities. Eur. J. Cancer **37**(4), 3–8 (2001)

Biological Structure Processing

Counting, Generating and Sampling Tree Alignments

Cedric Chauve[1], Julien Courtiel[1,2], and Yann Ponty[2,3](✉)

[1] Department of Mathematics, Simon Fraser University, Burnaby, Canada
{cedric.chauve,jcourtie}@sfu.ca
[2] Pacific Institute for the Mathematical Sciences, Vancouver, Canada
[3] CNRS-LIX, Ecole Polytechnique, Palaiseau, France
yann.ponty@lix.polytechnique.fr

Abstract. Pairwise ordered tree alignment are combinatorial objects that appear in RNA secondary structure comparison. However, the usual representation of tree alignments as supertrees is ambiguous, i.e. two distinct supertrees may induce identical sets of matches between identical pairs of trees. This ambiguity is uninformative, and detrimental to any probabilistic analysis. In this work, we consider tree alignments up to equivalence. Our first result is a precise asymptotic enumeration of tree alignments, obtained from a context-free grammar by means of basic analytic combinatorics. Our second result focuses on alignments between two given ordered trees. By refining our grammar to align specific trees, we obtain a decomposition scheme for the space of alignments, and use it to design an efficient dynamic programming algorithm for sampling alignments under the Gibbs-Boltzmann probability distribution. This generalizes existing tree alignment algorithms, and opens the door for a probabilistic analysis of the space of suboptimal RNA secondary structures alignments.

Keywords: Tree alignment · RNA secondary structure · Dynamic programming

1 Introduction

Tree alignments are the natural analog of sequence alignments, and have been introduced by Jiang et al. [10] to model and quantify the similarity between two (ordered[1]) trees. Initially proposed as an alternative to tree-edit distance, the tree alignment model has proven more robust, allowing for the inclusion of complex local operations [2], and for being generalized to multiple input trees [9]. Consequently, tree alignment has been used in a wide array of applicative contexts, especially RNA Bioinformatics [8], where RNA secondary structures alignments can be encoded by tree alignments. The minimal cost tree alignment between two trees of size n_1 and n_2, under classic insertion/deletion/(mis)-match

---

[1] In this work, unless explicitly specified, all trees will be rooted and ordered.

© Springer International Publishing Switzerland 2016
M. Botón-Fernández et al. (Eds.): AlCoB 2016, LNBI 9702, pp. 53–64, 2016.
DOI: 10.1007/978-3-319-38827-4_5

operations, can be computed using dynamic programming (DP). The current best algorithms have a worst-case time and space complexity respectively in $\mathcal{O}(n_1 n_2 (n_1 + n_2)^2)$ and $\mathcal{O}(n_1 n_2 (n_1 + n_2))$ [10] algorithms, and an average-case time and space complexity (on uniformly drawn instances) in $\mathcal{O}(n_1 n_2)$ [7].

In the context of sequence alignments, the enumeration of alignments has been the object of much interest in Computational Biology [1,5,13]. Alignments between two sequences over an alphabet Σ can be encoded as sequences over an extended alphabet Σ_a, representing insertions, deletions and (mis)matches (*e.g.* $\Sigma = \{a, b\}$, $\Sigma_a = \{(a, -), (-, b), (a, b), (a, a), (b, a), (b, b)\}$). Many sequences over Σ_a are equivalent if one considers only (mis)matches of the alignments, *i.e.* they align sequences of same length and induce the same sets of matched positions (*e.g.* $(a, -), (-, b)$ and $(-, b), (a, -)$). It is a natural problem to enumerate distinct sequence alignments for two sequences of cumulative length n [15, p. 188]. Beyond purely theoretical considerations, the decompositions introduced for enumerating distinct sequence alignments were adapted into DP algorithms, *e.g.* for probabilistic alignment based on expectation maximization [4], or to compute Gibbs-Boltzmann measures of reliability [14].

In the present work, we consider similar questions on *tree alignments*. We are first interested in counting distinct tree alignments, *i.e.* enumerating, up to equivalence, ordered trees whose vertices are labeled in Σ_a (called *supertrees* from now on). For trees, the notion of equivalence of alignments generalizes that of sequence alignments, *i.e.* two alignments are *equivalent* when they align the same pairs of trees, and induce the same sets of (mis)matched positions. Unfortunately, contrasting with the case of sequence alignments, existing DP algorithms for computing an optimal tree alignment [2,10,12] cannot be easily adapted into enumeration schemes for tree alignments up to equivalence. This additional difficulty is due to the existence of ambiguities of different nature.

Our main contribution is a grammar for (distinct) tree alignments, which provably generates a single representative for each equivalence class. We use the symbolic method [6] to obtain the generating function of tree alignments, and asymptotic equivalents for various statistics of interest can easily be derived, such as the average number of alignments over trees of total size n. Finally, and, perhaps more importantly from an applied point of view, the grammar can be transformed into an unambiguous and complete DP algorithm for aligning two input trees. The resulting algorithm has the same asymptotic worst-case and average-case complexities, up to reasonable constants, as the current best – ambiguous – algorithm [2,10]. The main interest of such an algorithm is that it opens immediately the way to new applications for the tree alignment model, including a critical assessment of the reliability of optimal alignments, either obtained by counting co-optimal alignments, or by sampling suboptimal alignments according to a Gibbs-Boltzmann distribution (see [11] for an example of this approach for the RNA folding problem).

In Sect. 2 we introduce the main definitions about trees, supertrees and tree alignments. In Sect. 3, we provide a grammar that generates all tree alignments. In Sect. 4.1 we analyze this grammar from an enumerative point of view and give precise results on the number of alignments of fixed size. Finally, in Sect. 4.2 we show how to transform the tree alignments grammar into a dynamic programming algorithm to sample tree alignments between two specified trees. A long version with proofs is available on arXiv or HAL [3].

2   Definitions

Trees and supertrees. Let Σ be an *alphabet*. A tree T on Σ is a rooted plane tree whose vertices are labeled by elements of Σ. We denote by V_T the set of vertices of T. We *remove a non-root vertex* v from a tree T by contracting the edge between v and its parent u, that keeps its label. Removing the root r of a tree consists in creating a forest composed of the subtrees rooted at the children of r. We denote the operation of removing a vertex v from T by $T - v$.

We denote by Σ_a the alphabet defined by $\Sigma_a = (\Sigma \cup \{-\})^2 - \{(-,-)\}$. An element $(x, y) \in \Sigma_a$ is an *insertion* (resp. *deletion, match*) if $y = -$ (resp. $x = -$, $(x, y) \in \Sigma^2$). A *supertree* A is a tree on Σ_a; a vertex of A is an insertion (resp. deletion, match) if its label is an insertion (resp. deletion, match). The size of a supertree A is the number of its insertions and deletions, plus twice the number of its matches. A *superforest* is an ordered sequence of supertrees.

Given a supertree A on Σ, we define two forests $\pi_1(A)$ and $\pi_2(A)$ as follows: $\pi_1(A)$ (resp. $\pi_2(A)$) is obtained by (1) iteratively removing all insertion (resp. deletions) of A, in an arbitrary order, and (2) replacing the label (x, y) of each remaining vertex by x (resp. y). We refer to Fig. 1 for an illustration. We extend the notations π_1 and π_2 on vertices: for a non-insertion (resp. non-deletion) vertex v of A, we denote by $\pi_1(v)$ (resp. $\pi_2(v)$) the corresponding vertex in $\pi_1(A)$ (resp. $\pi_2(A)$). A vertex x of $\pi_1(A)$ such that $\pi_1^{-1}(x)$ is an insertion (resp. match) is said to be inserted (resp. matched) in A. Similarly, a vertex y of $\pi_2(A)$ such that $\pi_2^{-1}(y)$ is a deletion (resp. match) is said to be deleted (resp. matched) in A.

Tree alignments. As forests $\pi_1(A)$ and $\pi_2(A)$ are embedded into the supertree A, the latter implicitly defines an *alignment* between the forests $\pi_1(A)$ and $\pi_2(A)$, *i.e.* a set of correspondences between vertices of $\pi_1(A)$ and $\pi_2(A)$, that is consistent with the structure of both forests [10]. We refer to Fig. 1 for an illustration.

We now turn to the central notion of *equivalent alignments*, *i.e.* alignments of identical pairs of trees, that contain exactly the same set of matched vertices. Given a supertree A, representing an alignment between two trees $S = \pi_1(A)$ and $T = \pi_2(A)$, the *set of matches of* A is formed by the elements (x, y) of $V_S \times V_T$ such that $\pi_1^{-1}(x) = \pi_2^{-1}(y)$ (*i.e.* there exists a vertex v of A such that $\pi_1(v) = x$ and $\pi_2(v) = y$). Two supertrees A_1 and A_2 are *equivalent* if $\pi_1(A_1) = \pi_1(A_2)$, $\pi_2(A_1) = \pi_2(A_2)$, and the sets of matches of A_1 and A_2 are identical (see Fig. 2 for an illustration).

A *tree alignment* is then defined as an equivalence class over supertrees with respect to the above-defined equivalence relation, for which $\pi_1(A)$ and $\pi_2(A)$ are trees. The notion of *forest alignment* is similarly defined when $\pi_1(A)$ and $\pi_2(A)$

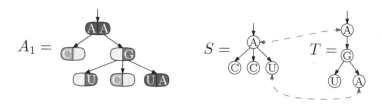

Fig. 1. A supertree A_1 with alphabet $\Sigma = \{A, C, G, U\}$, and the associated trees $S = \pi_1(A_1)$ and $T = \pi_2(A_1)$. The alignment of S and T defined by A is composed of two pairs of matched (A, A) and (U, A), indicated by dashed arrows.

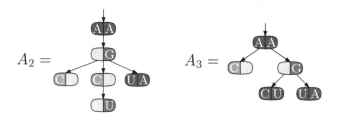

Fig. 2. Two non-equivalent supertrees, representing two different tree alignments. However, the supertree A_1 from Fig. 1 and the supertree A_2 are equivalent.

are not restricted to trees. Given a set $\mathcal{S}$ of tree (resp. forest) alignments, a set $\mathcal{T}$ of supertrees (resp. superforests) is said to be *representative of* $\mathcal{S}$ if it contains exactly one supertree (resp. superforest) for each alignment (*i.e.* equivalence classes of supertrees and forests) in $\mathcal{S}$. Tree alignments will now be the focus of our work.

3   A Grammar for Tree Alignments

In this section, we describe a context-free grammar for a set $\mathcal{A}$ of supertrees that is representative of the set of all tree alignments.

We first define some basic operations on supertrees and superforests:

– The (ordered) concatenation of two (super)forests A and B is denoted by $A \circ B$. It creates a new superforest beginning by the supertrees of A, and ending by the supertrees of B.
– Given two disjoint sets $\mathcal{T}_1$ and $\mathcal{T}_2$ of supertrees or superforests, we denote by $\mathcal{T}_1 \oplus \mathcal{T}_2$ their (disjoint) union.
– For any superforest A and $a, b \in \Sigma$, InsRoot (A, a) (resp. DelRoot (A, b), MatchRoot (A, a, b)) denotes the supertree whose root is the vertex $(a, -)$ (resp. $(-, b)$, (a, b)) and whose children are the supertrees in A, ordered with the same order that they have in A.

$$\mathcal{A} = \mathcal{V}^{\varnothing} \oplus \mathcal{T}_I \oplus \mathcal{T}_D \oplus \mathrm{InsRoot}\,(\mathcal{F}_I \circ \mathcal{T}_D) \tag{1}$$

$$\mathcal{T}_I = \mathrm{InsRoot}\,(\mathcal{F}_I), \quad \mathcal{F}_I = \{\text{empty superforest}\} \oplus \mathrm{InsRoot}\,(\mathcal{F}_I) \circ \mathcal{F}_I \tag{2}$$

$$\mathcal{T}_D = \mathrm{DelRoot}\,(\mathcal{F}_D), \quad \mathcal{F}_D = \{\text{empty superforest}\} \oplus \mathrm{DelRoot}\,(\mathcal{F}_D) \circ \mathcal{F}_D \tag{3}$$

$$\mathcal{V}^{\varnothing} = \mathcal{V}^{\uparrow} \oplus \mathrm{InsRoot}\,(\mathcal{V}\mathcal{H}) \tag{4}$$

$$\mathcal{V}^{\uparrow} = \mathrm{MatchRoot}\,(\mathcal{H}_{\mathsf{ID},\varnothing,\varnothing}) \oplus \mathrm{DelRoot}\left(\mathcal{F}_D \circ \mathcal{V}^{\uparrow} \circ \mathcal{F}_D\right) \tag{5}$$

$$\mathcal{V}\mathcal{H} = \mathcal{F}_I \circ \mathcal{V}\mathcal{H} \oplus \mathcal{V}^{\varnothing} \circ \mathcal{F}_I \oplus \mathrm{DelRoot}\,(\mathcal{H}_{\mathsf{ID},\mathsf{LR},\varnothing}) \circ \mathcal{F}_I \tag{6}$$

For every ν, M, M' with $\nu \in \{\mathsf{ID}, \mathsf{D}\}$ and $M, M' \in \{\varnothing, \mathsf{LR}, \mathsf{R}\}$:

$$\mathcal{H}_{\nu,M,M'} = \bigoplus \begin{cases} \{\text{empty superforest}\} & \text{if } (M, M') = (\varnothing, \varnothing) \\ \mathcal{T}_I \circ \mathcal{H}_{\nu,M,M'} & \text{if } \nu \neq \mathsf{D} \text{ and if } M \neq \mathsf{LR} \\ \mathcal{T}_D \circ \mathcal{H}_{\mathsf{D},M,M'} & \text{if } M' \neq \mathsf{LR} \\ \mathcal{V}^{\varnothing} \circ \overline{\mathcal{H}}_{M,M'}^{1,1} \\ \mathrm{InsRoot}\,(\mathcal{H}_{\mathsf{ID},\varnothing,\mathsf{LR}}) \circ \overline{\mathcal{H}}_{M,M'}^{1,+} \\ \mathrm{DelRoot}\,(\mathcal{H}_{\mathsf{D},\mathsf{LR},\varnothing}) \circ \overline{\mathcal{H}}_{M,M'}^{+,1} \end{cases} \tag{7}$$

For every $M, M' \in \{\varnothing, \mathsf{LR}, \mathsf{R}\}$ and $i, j \in \{\mathbf{1}, \mathbf{+}\}$:

$$\overline{\mathcal{H}}_{M,M'}^{i,j} = \mathcal{H}_{\mathsf{ID},\alpha(M),\alpha(M')} \oplus \begin{cases} \mathcal{F}_I & \text{if } M = \varnothing \text{ and } M' = \mathsf{R} \\ \mathcal{F}_I & \text{if } M = \varnothing, M' = \mathsf{LR} \text{ and } j = + \\ \mathcal{F}_D & \text{if } M = \mathsf{R} \text{ and } M' = \varnothing \\ \mathcal{F}_D & \text{if } M = \mathsf{LR}, M' = \varnothing \text{ and } i = + \\ \varnothing & \text{otherwise} \end{cases} \tag{8}$$

where $\alpha(\varnothing) = \varnothing$ and $\alpha(\mathsf{LR}) = \alpha(\mathsf{R}) = \mathsf{R}$.

Fig. 3. A context-free grammar for $\mathcal{A}$, a representative set of all tree alignments.

– We naturally extend these operators to a set $\mathcal{T}$ of supertrees or superforests:
$$\mathrm{InsRoot}\,(\mathcal{T}) = \bigoplus_{A \in \mathcal{T}, a \in \Sigma} \mathrm{InsRoot}\,(A, a), \quad \mathrm{DelRoot}\,(\mathcal{T}) = \bigoplus_{A \in \mathcal{T}, a \in \Sigma} \mathrm{DelRoot}\,(A, a),$$
$$\mathrm{MatchRoot}\,(\mathcal{T}) = \bigoplus_{A \in \mathcal{T}, (a,b) \in \Sigma^2} \mathrm{MatchRoot}\,(A, a, b).$$

Our grammar is described in Fig. 3, and illustrated in Fig. 4. The start symbol is $\mathcal{A}$ and the terminal states are the empty superforests of Eqs. (2), (3), (7).

Theorem 1. *The set of supertrees $\mathcal{A}$ generated by the grammar (1)–(8) is representative of the set of all tree alignments; i.e. $\mathcal{A}$ contains exactly one supertree for each equivalence class of supertrees.*

The key ingredient to prove Theorem 1 stems from the following (semantic) properties for the classes of supertrees and forests that appear in the grammar:

1. Supertrees in $\mathcal{T}_I$ (resp. $\mathcal{T}_D$) contain only insertion (resp. deletion) vertices.

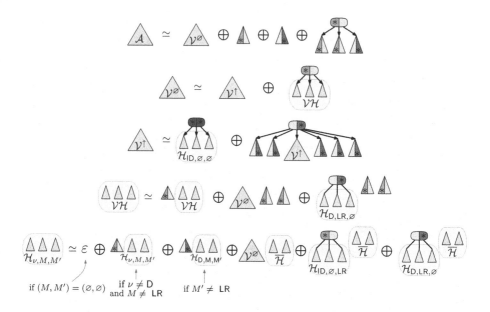

Fig. 4. A schematic illustration of the grammar for tree alignments.

2. $\mathcal{F}_I$ (resp. $\mathcal{F}_D$) is the set of superforests formed by supertrees of $\mathcal{T}_I$ (resp. $\mathcal{T}_D$).
3. For $\mu \in \{\varnothing, \uparrow\}$, $\mathcal{V}^\mu$ is representative of the set of alignments A with at least one match, such that, if $\mu = \uparrow$, then the root of $\pi_1(A)$ is matched.
4. $\mathcal{VH}$ is representative of the set of forest alignments A with at least one match, such that $\pi_2(A)$ is a tree.
5. For $\nu \in \{\mathsf{ID}, \mathsf{D}\}$ and $(M, M') \in \{\varnothing, \mathsf{LR}, \mathsf{R}\}^2$, $\mathcal{H}_{\nu, M, M'}$ is representative of the set of superforests A such that
   - if $\pi_1(A) \neq \varnothing$ and $\nu = \mathsf{D}$, then the first tree of $\pi_1(A)$ is matched in A;
   - if $M = \mathsf{R}$, then the last tree of $\pi_1(A)$ is matched in A (so $\pi_1(A) \neq \varnothing$);
   - if $M' = \mathsf{R}$, then the last tree of $\pi_2(A)$ is matched in A (so $\pi_2(A) \neq \varnothing$);
   - if $M = \mathsf{LR}$, then the first and last trees in $\pi_1(A)$ are matched in A (so $\pi_1(A)$ has at least two trees);
   - if $M' = \mathsf{LR}$, then the first and last trees in $\pi_2(A)$ are matched in A (so $\pi_2(A)$ has at least two trees).
6. For $i, j \in \{1, +\}^2$, $\overline{\mathcal{H}}_{M, M'}^{i,j}$ is representative of superforests A' such that
   - there exists a superforest A such that $A \circ A' \in \mathcal{H}_{\mathsf{D}, M, M'}$;
   - if $i = 1$ (resp. $+$), $\pi_1(A)$ is a tree (resp. a forest with at least two trees);
   - if $j = 1$ (resp. $+$), $\pi_2(A)$ is a tree (resp. a forest with at least two trees).

These properties can be verified recursively through a tedious analysis of the grammar, and imply quite straightforwardly that $\mathcal{A}$ contains one and exactly one supertree per equivalence class of supertrees.

Remark 1. *For sequences alignments, a grammar generating a representative set of sequence alignments can be easily adapted from the grammar generating* *all sequences over Σ_a, e.g. by preventing any occurrence of a deletion to imme-diately precede an insertion. Our grammar for tree alignments is constructed on the same principle: insertions are preferred to deletions. However, unlike sequences, the two-dimensional nature of the trees seems to forbid an explicit characterization of the representative elements, and seems to intrinsically man-date intricate combinatorial constructs/grammars. Note that our grammar, while complex, remains amenable to efficient computations (Sect. 4).*

4 Applications

4.1 Enumerating Tree Alignments

For the sake of simplicity, we will restrict our attention to $|\Sigma| = 1$, *i.e.* the alphabet is restricted to a single letter. The general case follows easily, and will be described in an extended version of the paper.

For a family $\mathcal{F}$ of superforests, we define a bivariate ordinary generating function

$$F(t, z) = \sum_{n \geq 0,\, k \geq 0} f_{n,k}\, t^n\, z^k$$

where $f_{n,k}$ is the number of superforests in $\mathcal{F}$ of size n with k matches.

Using the *symbolic method* [6], one classically translates the specification described by Eqs. (1)–(8) into a system of functional equations relating the gen-erating functions of the sets of supertrees and forests. To that purpose, classes of objects are replaced by their generating function, disjoint unions (resp. concate-nations) of two sets of supertrees are replaced by additions (resp. multiplications) of their generating functions, the addition of a root translates into a multipli-cation by a monomial tz (resp. t) if the root represents a match (resp. inser-tion/deletion), and empty superforests and sets translate into 1 and 0 respec-tively. The grammar is context-free, so the resulting system is algebraic and can be solved to yield the following characterization result.

Theorem 2. *The generating functions $T(t, z)$ and $F(t, z)$ of tree and forest alignments, whose size and number of matches are marked by t and z respec-tively, satisfy*

$$T(t, z) = \left(t^2 + t - t^2 z + \frac{t}{\sqrt{1 - 4t}} \right) F(t, z), \tag{9}$$

$$(tzC(t)^2 - t^2C(t)^2 + 2t)F(t, z)^2 + (t^2C(t)^4 - 2tC(t)^2 - 1)F(t, z) + C(t)^2 = 0, \tag{10}$$

where $C(t) = (1 - \sqrt{1 - 4t})/2t$ is the generating function of Catalan numbers.

Solving the quadratic Eq. (10) leads to an explicit formula for *FA* (and hence *TA*), details of which are omitted due to space constraints. Nonetheless, these explicit expressions can be used to compute an asymptotic estimate using a *transfer theorem* [6, Cor.VI.1 p. 392].

Theorem 3. *The number of tree alignments of size n is asymptotically equiva-lent to $\kappa \times n^{-3/2} \times 6^n$, where $\kappa = \sqrt{2}(3 - \sqrt{3})/(24\sqrt{\pi})$.*

Corollary 1. *The average number of tree alignments for a random pair of trees of cumulative size n is $\kappa' \times 1.5^n$, where $\kappa' = \sqrt{2}(3 - \sqrt{3})/6$.*

Similar techniques can be used to characterize the distribution of the number of matches in a random tree alignment. A direct application of [6, Theorem 9.12 p. 676] indeed gives the following.

Proposition 2. *Let m_n be the random variable that counts the number of matches in a uniformly-drawn random tree alignment. The variable m_n follows a Normal law of mean $\mathbb{E}(m_n) \sim n/6$ and variance $\mathbb{V}(m_n) \sim n/6$.*

4.2   Sampling Alignments Between Two Given Trees

We now consider two *fixed* trees S and T, and consider the task of sampling a tree alignment A such that $\pi_1(A) = S$ and $\pi_2(A) = T$, with respect to the Gibbs-Boltzmann probability distribution. This can be used to assess the stability of a prediction. We refer the interested reader to our introduction for examples of further motivation and possible applications.

Preliminaries. Let $\mathcal{T}_{S,T}$ be the set of all supertrees A such that $\pi_1(A) = S$ and $\pi_2(A) = T$, and $\mathcal{A}_{S,T}$ be a representative set of $\mathcal{T}_{S,T}$. In other words, $\mathcal{A}_{S,T}$ can be interpreted as the set of all alignments between S and T. For any supertree $A \in \mathcal{T}_{S,T}$, we define its *edit score* $s(A)$ as the sum of the number of insertions, deletions and mismatches (that is, matches (x, y) such that $x \neq y$).[2]

For a given positive constant $k\theta$, the *partition function* $Z_{S,T}$ of $\mathcal{A}_{S,T}$ and the *Gibbs-Boltzmann probability* $\Pr(A)$ of an alignment $A \in \mathcal{A}_{S,T}$ are defined as

$$Z_{S,T} = \sum_{A \in \mathcal{A}_{S,T}} e^{-s(A)/k\theta}, \quad \Pr(A) = \frac{e^{-s(A)/k\theta}}{Z_{S,T}}.$$

When $k\theta$ tends to 0, this distribution tends to the uniform distribution over supertrees of minimum edit score, while, when $k\theta$ tends to $+\infty$, it tends toward the uniform distribution over $\mathcal{A}_{S,T}$.

We consider the following problem: given two trees S and T, and a positive constant $k\theta$, design a sampling algorithm for alignments between S and T under the Gibbs-Boltzmann probability distribution. This problem generalizes the classic combinatorial optimization problem of computing a tree alignment between S and T having minimum edit score.

To address this problem, we rely on dynamic programming, by the approach described, among others, in [11] for RNA folding. We begin by adapting the grammar introduced in Sect. 3 into a grammar for $\mathcal{A}_{S,T}$, then detail how this grammar leads to an efficient sampling algorithm.

---

[2] The present results can be trivially extended to any edit scoring system that is a positive linear combination of the numbers of insertions, deletions and matches.

A grammar for $\mathcal{A}_{S,T}$. In order to guarantee that each supertree A indeed aligns two input trees S and T (namely $\pi_1(A) = S$ and $\pi_2(A) = T$), we need to restrict which rules in the grammar can be used, conditionally to which trees and forests are currently being generated. To that purpose, we introduce, for each set $\mathcal{S}$ in the previous grammar, an indexed version $\mathcal{S}_{[u,v]}$ which denotes the restriction of $\mathcal{S}$ to alignments between u and v two forests in S and T.

Slightly abusing previous notations, we denote by $a(u)$ the tree whose root is a vertex a and whose (forest of) children is u. Finally, for every tree/forest X, $\text{Ins}(X)$ (resp. $\text{Del}(X)$) represents the supertree/superforest obtained from X by inserting (resp. deleting) each of its elements. If X is empty, $\text{Ins}(X)$ and $\text{Del}(X)$ denote the empty superforest. The grammar for $\mathcal{A}_{S,T}$ is described in Fig. 5.

Theorem 4. *Let S and T be non-empty trees. The set of supertrees $\mathcal{A}_{S,T}$ generated by grammar (11)–(18) is representative of $\mathcal{T}_{S,T}$ the tree alignments between S and T.*

Applications to dynamic programming. The grammar defined by Eqs. (11)–(18) is a decomposition scheme for the alignments between S and T. It can easily be transformed into an algorithm for computing the partition function $Z_{S,T}$. Indeed, $Z_{S,T}$ is simply a weighted sum over all possible supertrees of $\mathcal{A}_{S,T}$, which is a set generated by the grammar. Now consider the image of the grammar as a set of numerical equations, obtained by syntactically replacing:

- The operators $(\oplus, \circ)$ with $(\sum, \times)$ respectively;
- The empty set $\varnothing$ with 0;
- Inserted/Deleted trees/forests $\text{Ins}(X)$ and $\text{Del}(X)$ with $e^{-|X|/k\theta}$,
- Match MatchRoot (V, a, a) events with V, $\forall a \in \Sigma$ and any expression V;
- Insertion InsRoot (V, a) events, deletion DelRoot (V, a) events, and mismatch MatchRoot (V, a, b) events with $e^{-1/k\theta} \times V$, $\forall a \neq b \in \Sigma$ and any V.

Theorem 4 immediately implies that the resulting set is a dynamic programming scheme that computes $Z_{S,T}$ instead of $\mathcal{A}_{S,T}$.

Moreover, each non-terminal term of the modified grammar now contains the partition function of the set of supertrees associated to this non-terminal term in the set-theoretic grammar, *e.g.* a term $\mathcal{VH}[a(u) \circ X, b(v)]$. This information can then be used to define an algorithm to sample supertrees from $\mathcal{A}_{S,T}$ under the Gibbs-Boltzmann distribution, following the *recursive* method for random generation [16].

To do so, it suffices to reinterpret the grammar defined by Eqs. (11)–(18) as a branching process: each $\oplus$ operator is replaced by a branching operator that, instead of joining sets of supertrees into a larger set of supertrees, chooses one of the sets according to the weight of its partition function. For instance, assume we have a grammar rule $U = V \oplus W$: the sampling algorithm will select one of the sets V, W, with V being chosen with probability $Z_V/(Z_V + Z_W)$, and W with probability $Z_W/(Z_V + Z_W)$, provided that Z_V, Z_W and Z_X have been previously computed. Recursive calls will then result into a supertree, which is provably randomly generated under the Gibbs-Boltzmann distribution.

$$\mathcal{A}_{\substack{S,T \\ S \equiv r_S(X_S)}} = \mathcal{V}^{\varnothing}[S,T] \oplus \text{InsRoot}\left(\text{Ins}(X_S) \circ \text{Del}(T), r_S\right) \tag{11}$$

$$\mathcal{V}^{\varnothing}[a(u), b(v)] = \mathcal{V}^{\uparrow}[a(u), b(v)] \oplus \text{InsRoot}\left(\mathcal{V}\mathcal{H}[u, b(v)], a\right) \tag{12}$$

$$\mathcal{V}^{\uparrow}[a(u), b(v)] = \bigoplus \begin{cases} \text{MatchRoot}\left(\mathcal{H}_{\text{ID},\varnothing,\varnothing}[u, v], a, b\right) \\ \displaystyle\bigoplus_{Y \circ c(w) \circ Y' = v} \text{DelRoot}\left(\text{Del}(Y) \circ \mathcal{V}^{\uparrow}[a(u), c(w)] \circ \text{Del}(Y'), b\right) \end{cases} \tag{13}$$

$$\mathcal{V}\mathcal{H}[\varnothing, b(v)] = \varnothing \tag{14}$$

$$\mathcal{V}\mathcal{H}[a(u) \circ X, b(v)] = \bigoplus \begin{cases} \text{Ins}(a(u)) \circ \mathcal{V}\mathcal{H}[X, b(v)] \\ \displaystyle\bigoplus_{\substack{X' \circ X'' = a(u) \circ X \\ |X'| \geq 2}} \text{DelRoot}\left(\mathcal{H}_{\text{ID},\text{LR},\varnothing}[X', v], b\right) \circ \text{Ins}(X'') \\ \mathcal{V}^{\varnothing}[a(u), b(v)] \circ \text{Ins}(X) \end{cases} \tag{15}$$

For every ν, M, M' with $\nu \in \{\text{ID}, \text{D}\}$ and $M, M' \in \{\varnothing, \text{LR}, \text{R}\}$:

$$\mathcal{H}_{\nu,M,M'}[X, \varnothing] = \begin{cases} \text{Ins}(X) & \text{if } (M, M') = (\varnothing, \varnothing), \\ \varnothing & \text{otherwise,} \end{cases} \tag{16}$$

$$\mathcal{H}_{\nu,M,M'}[\varnothing, Y] = \begin{cases} \text{Del}(Y) & \text{if } (M, M') = (\varnothing, \varnothing), \\ \varnothing & \text{otherwise,} \end{cases} \tag{17}$$

$\mathcal{H}_{\nu,M,M'}[a(u) \circ X, b(v) \circ Y] =$

$$\bigoplus \begin{cases} \text{Ins}(a(u)) \circ \mathcal{H}_{\nu,M,M'}[X, b(v) \circ Y] & \text{if } \nu \neq \text{D and if } M \neq \text{LR}, \\ \text{Del}(b(v)) \circ \mathcal{H}_{\text{D},M,M'}[a(u) \circ X, Y] & \text{if } M' \neq \text{LR}, \\ \mathcal{V}^{\varnothing}[a(u), b(v)] \circ \mathcal{H}_{\text{ID},\alpha(M,X),\alpha(M',Y)}[X, Y] \\ \displaystyle\bigoplus_{\substack{Y' \circ Y'' = b(v) \circ Y \\ |Y'| \geq 2}} \text{InsRoot}\left(\mathcal{H}_{\text{ID},\varnothing,\text{LR}}[u, Y'], a\right) \circ \mathcal{H}_{\text{ID},\alpha(M,X),\alpha(M',Y'')}[X, Y''] \\ \displaystyle\bigoplus_{\substack{X' \circ X'' = a(u) \circ X \\ |X'| \geq 2}} \text{DelRoot}\left(\mathcal{H}_{\text{D},\text{LR},\varnothing}[X', v], b\right) \circ \mathcal{H}_{\text{ID},\alpha(M,X''),\alpha(M',Y)}[X'', Y] \end{cases} \tag{18}$$

where $\alpha(\varnothing, X) = \varnothing$ and $\alpha(\text{LR}, X) = \alpha(\text{R}, X) = \begin{cases} \varnothing & \text{if } X = \varnothing, \\ \text{R} & \text{otherwise.} \end{cases}$

Fig. 5. A grammar for $\mathcal{A}_{S,T}$, a representative set of all tree alignments between two fixed trees S and T.

Theorem 5. *Let S and T be two trees of respective sizes n_S and n_T. The above-defined branching process adapted from grammar (11)–(18) defines an algorithm that samples a supertree from $\mathcal{A}_{S,T}$ under the Gibbs-Boltzmann distribution. The worst-case time and space complexities of the algorithm are in $\mathcal{O}(n_S\, n_T\, (n_S + n_T)^2)$, while the average-case time and space complexities are in $\mathcal{O}(n_S\, n_T)$.*

The correctness of the algorithm immediately follows from Theorem 4. Its complexities are identical to [7,10] since the structure of the DP scheme essentially remains the same; only the number of DP tables is increased (by a constant factor). This implies that our algorithm, while solving a much more general problem, retains the same asymptotic complexity (up to constants) than the current tree alignment algorithms that are limited to computing a single optimal tree alignment.

5   Conclusion and Discussion

Following a classical line of research in string algorithms, we introduced the notion of equivalence for tree alignments, and described a context-free grammar for a representative set of all possible alignments. We also showed how this grammar can be used to derive asymptotic properties of alignments, and design an efficient dynamic programming sampling algorithm for alignments between two given trees.

From an applied point of view, our results allow to sample optimal, as well as suboptimal, tree alignments for a pair of given trees under the Gibbs-Boltzmann distribution; following the program outlined in [11], we are currently using this algorithm to revisit the alignment of RNA structures.

Our proposed grammar for tree alignments is more complex than the grammars used to generate a representative set of sequence alignments, although dynamic programming for computing optimal sequences and trees alignments are very similar. This is due to the fact that it is particularly hard to characterize a representative set of tree alignments (see Remark 1). It thus remains an open problem to design a representative set of tree alignment that would be amenable to enumeration using a simpler grammar. However, it is important to remark that, despite its apparent complexity, our grammar leads to algorithms with an asymptotic complexity of the same order than existing optimization algorithms.

From a theoretical point of view, we believe that tree alignments as defined in this work form an interesting combinatorial family whose properties deserve to be explored in depth. More generally, it would be interesting to characterize the conditions under which an instance-agnostic grammar, enumerating a search space, could be adapted into a decomposition for a specific instance. Such a theory, at the confluence of enumerative combinatorics and algorithmic design, could provide another principled ways to design dynamic-programming algorithms.

References

1. Andrade, H., Area, I., Nieto, J.J., Torres, A.: The number of reduced alignments between two dna sequences. BMC Bioinformatics **15**, 94 (2014). http://dx.doi.org/10.1186/1471-2105-15-94
2. Blin, G., Denise, A., Dulucq, S., Herrbach, C., Touzet, H.: Alignments of RNA structures. IEEE/ACM Trans. Comput. Biol. Bioinform. **7**(2), 309–322 (2010). http://doi.acm.org/10.1145/1791396.1791409

3. Chauve, C., Courtiel, J., Ponty, Y.: Counting, generating and sampling tree alignments. In: ALCOB - 3rd International Conference on Algorithms for Computational Biology - 2016. Trujillo, Spain, Jun 2016. https://hal.inria.fr/hal-01154030

4. Do, C.B., Gross, S.S., Batzoglou, S.: CONTRAlign: discriminative training for protein sequence alignment. In: Apostolico, A., Guerra, C., Istrail, S., Pevzner, P.A., Waterman, M. (eds.) RECOMB 2006. LNCS (LNBI), vol. 3909, pp. 160–174. Springer, Heidelberg (2006)

5. Dress, A., Morgenstern, B., Stoye, J.: The number of standard and of effective multiple alignments. Appl. Math. Lett. 11(4), 43–49 (1998). http://www.sciencedirect.com/science/article/pii/S0893965998000548

6. Flajolet, P., Sedgewick, R.: Analytic combinatorics. Cambridge University Press, Cambridge (2009)

7. Herrbach, C., Denise, A., Dulucq, S.: Average complexity of the Jiang-Wang-Zhang pairwise tree alignment algorithm and of a RNA secondary structure alignment algorithm. Theor. Comput. Sci. 411(26–28), 2423–2432 (2010). http://dx.doi.org/10.1016/j.tcs.2010.01.014

8. Höchsmann, M., Töller, T., Giegerich, R., Kurtz, S.: Local similarity in RNA secondary structures. Proc. Ieee Comput. Soc. Bioinform Conf. 2, 159–168 (2003)

9. Höchsmann, M., Voss, B., Giegerich, R.: Pure multiple rna secondary structure alignments: a progressive profile approach. IEEE/ACM Trans. Comput. Biol. Bioinformatics 1(1), 53–62 (2004). http://dx.doi.org/10.1109/TCBB.2004.11

10. Jiang, T., Wang, L., Zhang, K.: Alignment of trees - an alternative to tree edit. Theor. Comput. Sci. 143(1), 137–148 (1995). http://dx.doi.org/10.1016/0304-3975(95)80029-9

11. Ponty, Y., Saule, C.: A combinatorial framework for designing (pseudoknotted) RNA algorithms. In: Przytycka, T.M., Sagot, M.-F. (eds.) WABI 2011. LNCS, vol. 6833, pp. 250–269. Springer, Heidelberg (2011). http://dx.doi.org/10.1007/978-3-642-23038-7_22

12. Schirmer, S., Giegerich, R.: Forest alignment with affine gaps and anchors, applied in RNA structure comparison. Theor. Comput. Sci. 483, 51–67 (2013). http://dx.doi.org/10.1016/j.tcs.2012.07.040

13. Torres, A., Cabada, A., Nieto, J.J.: An exact formula for the number of alignments between two DNA sequences. DNA Seq. 14(6), 427–430 (2003)

14. Vingron, M., Argos, P.: Determination of reliable regions in protein sequence alignments. Protein Eng. 3(7), 565–569 (1990). http://peds.oxfordjournals.org/content/3/7/565.abstract

15. Waterman, M.S.: Introduction to Computational Biology: Maps, Sequences, and Genomes. CRC Press, Pevzner (1995)

16. Wilf, H.S.: A unified setting for sequencing, ranking, and selection algorithms for combinatorial objects. Adv. Math. 24, 281–291 (1977)

A New Multi-objective Approach for Molecular Docking Based on RMSD and Binding Energy

Esteban López-Camacho$^{(\boxtimes)}$, María Jesús García-Godoy, José García-Nieto, Antonio J. Nebro, and José F. Aldana-Montes

Khaos Research Group, Department of Computer Science, University of Málaga, ETSI Informática, Campus de Teatinos, 29071 Málaga, Spain
{esteban,mjgarciag,jnieto,antonio,jfam}@lcc.uma.es

Abstract. Ligand-protein docking is an optimization problem based on predicting the position of a ligand with the lowest binding energy in the active site of the receptor. Molecular docking problems are traditionally tackled with single-objective, as well as with multi-objective approaches, to minimize the binding energy. In this paper, we propose a novel multi-objective formulation that considers: the Root Mean Square Deviation (RMSD) difference in the coordinates of ligands and the binding (intermolecular) energy, as two objectives to evaluate the quality of the ligand-protein interactions. To determine the kind of Pareto front approximations that can be obtained, we have selected a set of representative multi-objective algorithms such as NSGA-II, SMPSO, GDE3, and MOEA/D. Their performances have been assessed by applying two main quality indicators intended to measure convergence and diversity of the fronts. In addition, a comparison with LGA, a reference single-objective evolutionary algorithm for molecular docking (AutoDock) is carried out. In general, SMPSO shows the best overall results in terms of energy and RMSD (value lower than 2Å for successful docking results). This new multi-objective approach shows an improvement over the ligand-protein docking predictions that could be promising in *in silico* docking studies to select new anticancer compounds for therapeutic targets that are multidrug resistant.

Keywords: Molecular docking · Multi-objective optimization · Nature inspired metaheuristics · Algorithm comparison

1 Introduction

Ligand-protein docking is an optimization problem which aims at predicting the position of a small molecule (ligand) to a receptor (macromolecule) with the goal of finding the ligand position to the receptor with a minimum binding energy. Molecular docking problem has been tackled with single-objective algorithms, to minimize the binding energy [11], as well as with multi-objective approaches, to minimize the intermolecular energy E_{inter} (energy interaction between ligand and the target) and the intramolecular energy E_{intra} (the internal energy compound) [4].

© Springer International Publishing Switzerland 2016
M. Botón-Fernández et al. (Eds.): AlCoB 2016, LNBI 9702, pp. 65–77, 2016.
DOI: 10.1007/978-3-319-38827-4_6

In this regard, a number of studies based on the application of multi-objective algorithms to the ligand-protein docking have been proposed. A first attempt was carried out in 2006 by Oduguwa et al. [15], in which three evolutionary multi-objective algorithms (NSGA-II, PAES, and SPEA) were applied to evaluate three objectives such as the E_{inter}, E_{intra} and shape complementarities on three molecular complexes. Grosdidier et al. [5] proposed a new hybrid evolutionary algorithm called EADock that optimizes two different energy score functions that evaluate the E_{inter}, E_{intra} and the solvation free energy. In 2008, Janson et al. [7] designed a parallel multi-objective algorithm using AutoDock 3.05 energy function, called ClustMPSO, minimizing as objectives the E_{inter} and E_{intra} when dealing with six molecular complexes. In the same year, Boisson et al. [1] implemented a parallel evolutionary bi-objective model based on optimizing two objectives: the sum of E_{inter} and E_{intra} and a surface term for the docking of six instances. Sandoval-Perez et al. [16] used the implementation of NSGA-II provided by the jMetal framework to optimize bound and non-bound energy terms of ligand/receptor as objectives applied to four docking instances. Gu et al. [6] developed a new multi-objective approach based on optimizing the solutions generated by an aggregated scoring function that includes terms from force-field, empirical and knowledge-based scoring functions.

In all these previous publications, a series of different multi-objectives formulations have been proposed that focus on energy scoring function. However, they do not consider guiding the search with a new objective when the co-crystallized ligand is known, which could complement the traditional energy function.

With this motivation, we propose in this work a novel multi-objective approach consisting minimizing: (1) the binding energy (the unbound and bound energy terms of the ligand/receptor complex), and (2) the Root-Mean-Square-Deviation (RMSD) score, when the co-crystallized ligand pose is known. These two main objectives have been used to evaluate the quality of the ligand-protein interactions. With this aim, we compare and analyze the performance of four multi-objective metaheuristics when solving 11 flexible ligand-receptor docking complexes taken from the AutoDock 4.2 benchmark [12]. This dataset includes flexible ligands with different sizes and flexible side-chains of HIV-protease receptors for more realistic results. The algorithms used in this study are: Nondominated Sorting Genetic Algorithm II (NSGA-II) [2], Speed Modulation Multi-Objective Particle Swarm Optimization (SMPSO) [13], Third Evolution Step of Generalized Differential Evolution (GDE3) [8], and Multi-Objective Evolutionary Algorithm Based on Decomposition (MOEA/D) [18]. These algorithms constitute a varied set of evolutionary and difference-vector multi-objective techniques representative of the state of the art, performing different learning procedures and inducing different behaviors in terms of convergence and diversity.

This paper is organized as follows: Sect. 2 describes the molecular docking problem from a multi-objective formulation. Studied algorithms are briefly described in Sect. 3. Section 4 reports the experimentation methodology and Sect. 5 analyzes the results obtained. Finally, Sect. 6 contains concluding remarks and future lines of research.

2 The Problem: Multi-objective Docking

A multi-objective optimization problem is characterized by two spaces: the decision and the objective spaces. The former involves all the possible feasible solutions, and the latter includes their corresponding objective values.

Decision space. The main objective in the molecular docking problem is to find an optimized conformation between the ligand (L) and the receptor (R) that results in the lowest binding energy. The ligand-receptor interaction is evaluated by an energy function calculated through three components representing degrees of freedom: (1) the translation of the ligand molecule, involving the three axis values (x, y, z) in cartesian coordinate space; (2) the ligand orientation, modeled as a four variables quaternion including the angle slope (θ); and (3) the flexibilities, represented by the free rotation of torsion (dihedral angles) of the ligand and sidechains of the receptor. Each problem solution for AutoDock and jMetal (the tools we have used) is encoded by a real-value vector of $7 + n$ variables, in which the first three values correspond to the ligand translation, the next four values correspond to the ligand and/or macromolecule orientation, and the remaining n values are the ligand torsion dihedral angles. Furthermore, in order to allow a rapid evaluation of the energy conformations, a grid-based methodology is implemented. The energy interaction is calculated and assigned to each grid point and is evaluated to obtain the energy of a given ligand pose [12].

Objective space. Our bi-objective formulation consists of: the E_{inter} and the RMSD score. The E_{inter} is the energy function as used in Autodock, that is calculated as follows:

$$E_{inter} = Q_{bound}^{R-L} + Q_{unbound}^{R-L} \qquad (1)$$

Q_{bound}^{R-L} and Q_{bound}^{R-L} are the states of bound and unbound of the ligand-receptor complex, respectively.

$$Q = W_{vdw} \sum_{i,j} (\frac{A_{ij}}{r_{ij}^{12}} - \frac{B_{ij}}{r_{ij}^{6}}) + W_{hbond} \sum_{i,j} E(t) \left(\frac{C_{ij}}{r_{ij}^{12}} - \frac{D_{ij}}{r_{ij}^{10}} \right)$$
$$+ W_{elec} \sum_{i,j} \frac{q_i q_j}{\varepsilon(r_{ij}) r_{ij}} + W_{sol} \sum_{i,j} (S_i V_j + S_j V_i) e^{(-r_{ij}^2/2\sigma^2)} \qquad (2)$$

Each pair of energetic evaluation terms includes evaluations (Q) of dispersion/repulsion (vdw), hydrogen bonds ($hbond$), electrostatics ($elec$) and desolvation (sol). Weights W_{vdw}, W_{hbond}, W_{conf}, W_{elec}, and W_{sol} of Eq. 2 are constants for Van der Waals, hydrogen bonds, torsional forces, electrostatic interactions and desolvation, respectively. r_{ij} represents the interatomic distance, A_{ij} and B_{ij} in the first term are Lennard-Jones parameters taken from the Amber force field. Similarly, C_{ij} and D_{ij} in the second term are Lennard-Jones parameters for maximum well depth of potential energies between two atoms, and $E(t)$ represents the angle-dependent directionality. The third term uses a Coulomb approach for electrostatics. Finally, the fourth term is calculated from the volume (V) of the atoms that are surrounding a given atom weighted by S, and an exponential term which involves atom distances. An extended explanation of all these variables can be found in [12].

The RMSD is a measure of similarity between the real ligand position in the receptor and the computed position of the docking ligand, that takes into account symmetry, partial symmetry (e.g. symmetry within a rotatable branch) and near-symmetry in a simple heuristic way. Ideally, the lower RMSD score the better solution is. A ligand-receptor docking solution with a RMSD score below 2Å is considered as a solution with high docking accuracy. It is worth noting that other docking solutions can be returned with higher RMSD scores and low values of E_{inter}, indicating that other possible interaction ligand sites should be considered. The RMSD score for two identical structures a and b is defined as follows:

$$RMSD_{ab} = max(RMSD'_{ab}, RMSD'_{ba}), \ with \ RMSD'_{ab} = \sqrt{\frac{1}{N} \sum_i \min_j r_2^{ij}}$$

(3)

The sum is over all N heavy atoms in structure a, the minimum is over all atoms in structure a with the same element type as atom i in structure b.

3   Algorithms

We have included in our study four algorithms which are representative of the state-of-the-art in the multi-objective optimization field. A brief description of each one of them is given next:

NSGA-II: NSGA-II [2] is a generational genetic algorithm, which uses the typical genetic operators (selection, crossover and mutation) to obtain new individuals from the original population. To promote convergence, a non-dominated sorting procedure based on Pareto ranking is used, while the crowding distance density estimator is applied to foster the diversity of the set of found solutions.

GDE3: The Generalized Differential Evolution (GDE) algorithm [8] is based on NSGA-II, but the genetic mutation and selection operators are replaced by their differential evolution counterparts. Furthermore, GDE3 modifies the crowding distance of NSGA-II as well to generate a better distributed set of solutions.

SMPSO: SMPSO [13] is a multi-objective particle swarm optimization algorithm. Its main feature is the limitation of the particle speed to allow new effective particle positions to be produced when the speed becomes too high. SMPSO uses the polynomial mutation as the turbulence factor and an external archive that stores the non-dominated solutions found during the search.

MOEA/D: MOEA/D [18] has become the typical representative decomposition-based multi-objective algorithm, where a multi-objective problem is decomposed into a set of single-objective subproblems that then optimized simultaneously. In this study we have used the variant MOEA/D-DE [9], which applies differential evolution as variation operators. This algorithm also applies a polynomial mutation operator to improve its search capability.

In short, we have selected the most widely used algorithm in the field (NSGA-II), a solver based in differential evolution (GDE3), a PSO (SMPSO) and an algorithm based on decomposition (MOEA/D).

4   Experimentation

In this section, we include the selected benchmark problems, the experimentation methodology we have followed, and the parameter settings of the algorithms.

4.1   Benchmark Problems

In this study, we have selected a benchmark composed of 11 complexes having receptor and ligand flexibility. The selection of these complexes has been motivated as they are actually difficult docking problems containing a wide range of ligand sizes (from small to large inhibitors). The receptors of these complexes have a tunnel-shaped active site that wraps around a peptidomimetic inhibitor [12]. The receptor is a dimer whose subunits are bridged by an arginine-aspartate salt bridge at the end of the tunnel. The docking studies performed with these instances in [12] to test the energy function of AutoDock 4.2 demonstrated that the most difficult problems are those which involve smaller ligands. This is due to the flexibility added to the receptor side-chains (ARG-8) that increases the space of ligand interaction. These instances have been taken from the PDB database[1] and they have been properly prepared for the docking simulations.'

Table 1 summarizes the set of problems selected showing the PDB accession code, the X-ray crystal structures names and the structure resolution (Å). For all instances, the torsional degrees of freedom (flexibility) for ligands and macromolecules are 10 and 6, respectively, selecting those torsions that allow the fewest number of atoms to move around the ligand core. Therefore, the total number of solution variables (n) is 23 (3 for translation, 4 for rotation quaternion, and 16 for torsional degrees).

4.2   Methodology

For this work, we have carried out a thorough experimentation consisting in performing 30 independent runs for each combination of algorithm and molecular instance. From these executions, we have calculated the median and interquartile range (IQR) as measures of central tendency and statistical dispersion,

---

[1] In URL: http://www.rcsb.org/pdb/home/home.do.

Table 1. The accession codes, the X-ray crystal structure and resolution taken from PDB database are presented.

PDB Code	Protein-ligand complexes	Resolution (Å)
1AJV	HIV-1 protease/AHA006	2.00
1AJX	HIV-1 protease/AHA001	2.00
1BV9	HIV-1 protease/α-D-glucose	2.20
1D4K	HIV-1 protease/Macrocyclic peptidomimetic inhibitor 8	1.85
1G2K	HIV-1 protease/AHA047	1.95
1HIV	HIV-1 protease/U75875	2.00
1HPX	HIV-1 protease/KNI-272	2.00
1HTF	HIV-1 protease/GR126045	2.20
1HTG	HIV-1 protease/GR137615	2.00
1HVH	HIV-1 protease/Q8261	1.80
2UPJ	HIV-1 protease/U100313	3.00

respectively. We have considered two quality indicators to assess the algorithm performance: Hypervolume (I_{HV}) and Unary Additive Epsilon Indicator ($I_{\epsilon+}$) [3]. The first indicator takes into account both convergence and diversity, whereas the second one ($I_{\epsilon+}$) gives a measure of the convergence degree of the obtained Pareto front approximations. In this sense, it is worth noting that we are dealing with a real-world optimization problem, and therefore the true Pareto fronts to calculate these two metrics are not known. To cope with this issue, we have generated a reference Pareto front for each instance by combining all the non-dominated solutions computed in all the executions of all the algorithms.

As mentioned, we have used the implementation of the four algorithms studied provided in the jMetalCpp framework [10] in combination with AutoDock 4.2 to evaluate the new generated solutions. To cope with the high computational requirements needed by carry out our experiments, we have used the Condor[2] system, a middleware platform managing close to 400 cores that acts as a distributed task scheduler (each task dealing with one independent run).

4.3   Parameter Setup

The selected algorithms have been configured with a population size of 150 individuals (particles in the case of SMPSO). The stopping condition has been set to compute a number of 1,500,000 function evaluations. These values were chosen because they are the default settings used by AutoDock and they have been used in other studies [14].

Each algorithm has been configured using the parameter setup recommended in the research study where it was proposed, and these parameters are used as

---

[2] In URL: http://research.cs.wisc.edu/htcondor/.

default in the jMetal framework. In particular, SBX crossover and polynomial mutation are the variation operators used in NSGA-II. The distribution indexes for both operators are $\eta_c = 20$ for crossover, and $\eta_m = 20$ for mutation. The crossover probability is $p_c = 0.9$ and the mutation probability is $p_m = 1/n$, being n the number of decision variables of the tackled problem. NSGA-II applies binary tournament selection. In the case of GDE3 (variant $rand/1/bin$), the two DE control parameters μ and C_r take a value of 0.5, whereas in MOEA/D μ is set to 0.5 and C_r is set to 1.0. Both MOEA/D and SMPSO use the polynomial mutation with the same settings applied in NSGA-II. In SMPSO, the acceleration coefficients φ_1 and φ_2 are set to 1.5, the inertia weight is $W = 0.9$, and the polynomial mutation is applied to one sixth of the particles in the swarm.

5 Results

This section is devoted to presenting and analyzing the results obtained in our study. We start by assessing the performance of the algorithms and then they are compared with the values of a single-objective approach.

5.1 Performance Comparisons

We start our analysis by discussing the results yielded by applying the I_{HV} indicator. Let us remind that this indicator is the sum of the contributed volume of each point of a front in respect to a reference point, and the higher the convergence and diversity degrees of a front, the higher its I_{HV} value. Table 2 shows the median and interquartile range of the computed solutions for I_{HV} quality indicators for the set of 11 docking instances and the four algorithms being compared. According to these results, SMPSO achieves the best I_{HV} values in all the eleven considered problems and MOEA/D is the second best performing technique. We have to note that many cells have a I_{HV} value equal to zero; this happens when all the points of the produced fronts are beyond the limits of the reference point. This happens in most of the problems in all the algorithms excepting SMPSO, which indicates we are facing a very hard optimization problem.

Table 2. Median and interquartile range of I_{HV} for each algorithm and instance. Best and second best median results have dark and light gray backgrounds, respectively.

	NSGAII	SMPSO	GDE3	MOEAD
1AJV	$0.00e+00_{0.0e+00}$	$3.51e-01_{4.0e-02}$	$0.00e+00_{0.0e+00}$	$0.00e+00_{2.9e-01}$
1AJX	$0.00e+00_{0.0e+00}$	$5.52e-01_{2.0e-02}$	$0.00e+00_{0.0e+00}$	$7.47e-03_{6.8e-01}$
1D4K	$0.00e+00_{0.0e+00}$	$4.93e-01_{1.3e-01}$	$0.00e+00_{0.0e+00}$	$0.00e+00_{0.0e+00}$
1G2K	$0.00e+00_{0.0e+00}$	$3.32e-01_{3.3e-02}$	$0.00e+00_{0.0e+00}$	$0.00e+00_{4.1e-01}$
1HIV	$0.00e+00_{0.0e+00}$	$5.96e-01_{1.3e-01}$	$0.00e+00_{0.0e+00}$	$0.00e+00_{0.0e+00}$
1HPX	$0.00e+00_{0.0e+00}$	$2.04e-01_{1.8e-01}$	$1.27e-01_{6.5e-01}$	$0.00e+00_{1.1e-01}$
1HTF	$0.00e+00_{0.0e+00}$	$5.26e-02_{1.3e-01}$	$0.00e+00_{4.6e-03}$	$2.78e-02_{3.3e-01}$
1HTG	$0.00e+00_{0.0e+00}$	$5.31e-02_{5.5e-02}$	$0.00e+00_{0.0e+00}$	$0.00e+00_{1.9e-01}$
1HVH	$0.00e+00_{0.0e+00}$	$7.67e-01_{3.7e-02}$	$0.00e+00_{0.0e+00}$	$5.31e-01_{7.7e-01}$
1VB9	$0.00e+00_{0.0e+00}$	$7.34e-01_{6.5e-02}$	$0.00e+00_{0.0e+00}$	$0.00e+00_{1.4e-01}$
2UPJ	$0.00e+00_{0.0e+00}$	$5.86e-01_{9.8e-02}$	$0.00e+00_{0.0e+00}$	$1.90e-01_{5.8e-01}$

Table 3. Median and interquartile range of $I_{\epsilon+}$ for each algorithm and instance. Best and second best median results have dark and light gray backgrounds, respectively.

	NSGAII	SMPSO	GDE3	MOEAD
1AJV	$5.23e+00_{1.2e+00}$	$5.60e-01_{9.8e-02}$	$5.00e+00_{1.0e+00}$	$3.87e+00_{4.4e+00}$
1AJX	$3.43e+00_{2.4e+00}$	$2.61e-01_{7.2e-02}$	$1.49e+00_{3.3e-01}$	$1.01e+01_{2.0e+00}$
1D4K	$8.06e+00_{2.7e+00}$	$4.56e-01_{1.4e-01}$	$8.56e+00_{5.7e-01}$	$4.65e+00_{2.8e+00}$
1G2K	$4.28e+00_{1.4e+00}$	$5.71e-01_{1.2e-01}$	$3.93e+00_{1.3e+00}$	$2.69e+00_{3.6e+00}$
1HIV	$5.12e+00_{1.2e+00}$	$2.63e-01_{2.1e-01}$	$4.69e+00_{1.4e+00}$	$4.07e+00_{1.6e+00}$
1HPX	$1.42e+01_{3.6e+00}$	$6.32e-01_{2.8e-01}$	$6.71e-01_{1.1e+01}$	$1.03e+01_{1.3e+01}$
1HTF	$1.76e+00_{5.5e-01}$	$9.30e-01_{3.0e-01}$	$1.13e+00_{8.0e-01}$	$7.94e-01_{9.2e-01}$
1HTG	$7.48e+00_{7.1e-01}$	$9.63e-01_{6.6e-02}$	$6.82e+00_{8.7e-01}$	$5.03e+00_{6.4e+00}$
1HVH	$5.94e+00_{1.5e+00}$	$1.34e-02_{2.7e-02}$	$4.93e+00_{1.7e+00}$	$4.16e-01_{2.1e+00}$
1VB9	$8.59e+00_{2.4e+00}$	$1.33e-01_{5.6e-02}$	$7.85e+00_{1.3e+00}$	$7.04e+00_{4.9e+00}$
2UPJ	$3.42e+00_{2.4e+00}$	$3.03e-01_{6.6e-02}$	$3.56e+00_{1.1e+00}$	$7.64e-01_{2.7e+00}$

Table 4. Average Friedman's rankings with Holm's Adjusted p-values ($\alpha = 0.05$) of compared algorithms (SMPSO, GDE3, MOEA/D, and NSGA-II) for the test set of 11 docking instances. Symbol * indicates the control algorithm and column at right contains the overall ranking of positions with regards to I_{HV} and $I_{\epsilon+}$.

Hypervolume (HV)			Epsilon ($I_{\epsilon+}$)		
Algorithm	Fri_{Rank}	$Holm_{Ap}$	Algorithm	Fri_{Rank}	$Holm_{Ap}$
*SMPSO	1.02	-	*SMPSO	1.09	-
MOEA/D	2.68	2.24e-03	MOEA/D	2.00	9.87e-02
GDE3	3.09	1.45e-04	GDE3	3.09	2.79e-04
NSGA-II	3.22	5.21e-05	NSGA-II	3.81	7.25e-07

A similar behavior can be observed in Table 3 with regards to $I_{\epsilon+}$, which is a convergence measure. According to these results, SMPSO obtains the best $I_{\epsilon+}$ values in ten out of the eleven problems, while MOEA/D gets the best value in one problem (1HTF) and the second best in all of them but 1HPX.

In order to provide these results with statistical meaning (in this study $\alpha = 0.05$), non-parametric statistical tests have been applied because in several cases the distributions of results did not follow the conditions of normality and homoscedasticity [17]. Therefore, the analyses and comparisons focus on the entire distribution of each of the two metrics studied. Specifically, we have applied Friedman's ranking and Holm's post-hoc multicompare tests [17] to know which algorithms are statistically worse than the control one (with the best ranking).

In this regard, as shown in Table 4, SMPSO is the best ranked technique according to I_{HV} (with a value of 1.02), followed by MOEA/D, GDE3, and NSGA-II. Therefore, SMPSO is established as the control algorithm for I_{HV} in the post-hoc Holm test, which is compared with the remaining algorithms. The adjusted p-values ($Holm_{Ap}$ in Table 4) resulting from these comparisons are, for the last three algorithms (MOEA/D, GDE3, and NSGA-II), lower than the confidence level, meaning that SMPSO is statistically better than these algorithms.

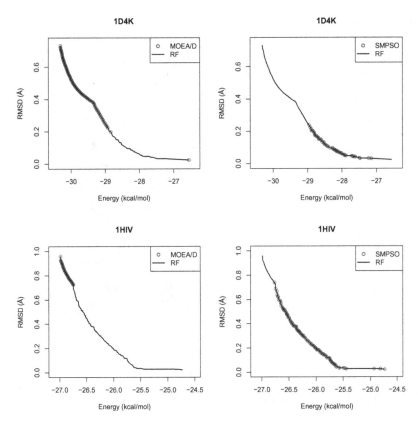

Fig. 1. Reference front contributions of docking instances 1D4K and 1HIV. SMPSO and MOEA/D contribute with practically all the solutions of the reference fronts.

In the case of $I_{\epsilon+}$, SMPSO is better ranked than the remaining compared algorithms, although without statistical differences in the case of MOEA/D. SMPSO is statistically better than GDE3 and NSGA-II.

In summary, SMPSO shows the overall best balance for the two quality indicators, followed by MOEA/D. These results are graphically supported by two examples included in Fig. 1, where the reference fronts obtained for two representative instances 1D4K and 1HIV are plotted. In these graphs, the contributions, in terms of solutions, of each algorithm to the global reference front are plotted with different points and colors. As it is easily observable, SMPSO and MOEA/D contribute with almost all solutions taking part to the reference front. Interestingly, SMPSO converges to the region biased towards the RMSD objective, whereas MOEA/D generate non-dominated solutions in a different region to the ones of SMPSO, thereby giving cue to the energy optimization. We can state that the specific learning procedures induced by SMPSO and MOEA/D lead these algorithms to search in different regions of the problem landscape, hence generating solutions in complementary parts of the reference front.

Table 5. Best RMSD scores (Å) calculated from all SMPSO solutions in comparison with the best RMSD values of LGA single-objective solutions.

	1AJV	1AJX	1BV9	1D4K	1G2K	1HIV	1HPX	1HTF	1HTG	1HVH	2UPJ
SMPSO	0.02	0.03	0.02	0.03	0.02	0.02	0.03	0.03	0.00	0.11	0.03
LGA	5.23	5.12	4.46	4.46	4.46	6.47	6.09	3.89	0.38	6.71	4.55

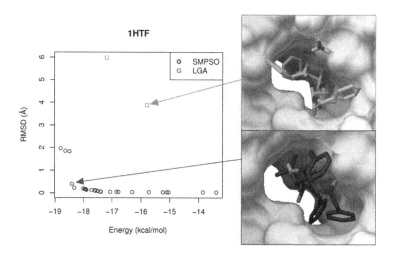

Fig. 2. Set of non-dominated solutions obtained by SMPSO with regards to those by LGA, for instance 1HTF. Corresponding ligand conformation structures (captured from AutoDock) of two representative solutions are shown at right side.

5.2   Comparison Single Versus Multi-objective

After the performance comparison of multi-objective algorithms, we are now interested in knowing how competitive their solutions are against those yielded by the LGA single-objective technique provided by Autodock 4.2. This way, we will be able to determine whether our bi-objective formulation has a positive effect in the search of solutions with the lowest RMSD score, or not. The values are included in Table 5, in which we can observe that SMPSO outperforms LGA in all the instance problems with large and small ligands.

With the multi-objective approach proposed here, SMPSO is able to return better results (RMSD scores below 2Å) for all the instances, since the optimization procedure is actively guided to compute solutions to the real ligand pose.

The use of RMSD as objective to guide the search procedure could be counter-intuitive, since it would restrict us to work only with molecular structures whose co-crystallized ligand are known beforehand (experimentally determined). Nevertheless, this new focus is useful in those typical cases in which the active site of a given therapeutic target mutates and makes it multidrug resistant. Therefore, new compounds analogous to the reference ligand should be tested to be considered as new pharmacological candidates.

Figure 2 shows the non-dominated solutions obtained by SMPSO, for molecule 1HTF. In addition, the two solutions with the best binding energy and RMSD values obtained by LGA are also plotted. In order to visualize the computed ligand docked to the active site of the HIV-protease receptor, we have selected two solutions with the best RMSD values. The best RMSD solutions for the LGA and the SMPSO are 3.89Å and 0.39Å with binding energies of −5.99 kcal/mol and −18.04 kcal/mol, respectively. As Fig. 2 shows, the ligand computed by the SMPSO is docked to the active site of the receptor with a better pose than the LGA computed ligand that is partially inside.

6 Conclusions

In this paper, we propose a novel multi-objective formulation of the molecular docking problem, where the RMSD and binding energy are the goals to optimize. This new approach has been incorporated in four multi-objective algorithms: NSGA-II, SMPSO, GDE3 and MOEA/D. A heterogenous set of 11 protein-ligand complexes with flexible ligands and receptors were selected in order to carry out the experiments. The main conclusions can be outlined as follows:

1. Using a multi-objective approach to solve the molecular docking could lead to a broad set of solutions, which can be selected according to the weight of the RMSD and binding energy, instead of only focusing on energy values.
2. SMPSO provides the best overall performance according to the two quality indicators used, and for the studied molecular instances.
3. For all studied molecular instances, SMPSO converges to the region biased towards the RMSD, whereas MOEA/D generates its fronts of non-dominated solutions in a different region, thereby giving cue to energy optimization.
4. According to the single-objective (AutoDock 4.2) fitness function, SMPSO algorithm find, in most of the cases, better solutions than the ones obtained by LGA. This is a noticeable result since SMPSO is a general purpose optimization technique, while LGA is specifically adapted to deal with the molecular docking problem.
5. The use of RMSD as objective to guide the search is useful in those typical cases in which the active site of a given therapeutic target mutates and makes it multi-drug resistant.

As future work, the most natural extension would be to design a hybrid algorithm combining search procedures from both SMPSO and MOEA/D algorithms in order to get solutions covering the full Pareto front. To test this, a greater number of molecular instances could be used and the solutions obtained could be studied from a more biological point of view.

Acknowledgments. This work is partially funded by Grants TIN2011-25840 (Ministerio de Ciencia e Innovación) and P11-TIC-7529 and P12-TIC-1519 (Plan Andaluz de Investigación, Desarrollo e Innovación). This article is based upon work from COST Action CA15140, supported by COST (European Cooperation in Science and Technology).

References

1. Boisson, J.C., Jourdan, L., Talbi, E., Horvath, D.: Parallel multi-objective algorithms for the molecular docking problem. In: IEEE Symposium on Computational Intelligence in Bioinformatics and Computational Biology, pp. 187–194, September 2008
2. Deb, K., Pratap, A., Agarwal, S., Meyarivan, T.: A fast and elitist multiobjective genetic algorithm: NSGA-II. IEEE Trans. Evol. Comput. **6**(2), 182–197 (2002)
3. Deb, K.: Multi-objective Optimization Using Evolutionary Algorithms. Wiley, New York (2001)
4. García-Godoy, M.J., López-Camacho, E., García-Nieto, J., Nebro, A.J., Aldana-Montes, J.F.: Solving molecular docking problems with multi-objective metaheuristics. Molecules **20**(6), 10154–10183 (2015)
5. Grosdidier, A., Zoete, V., Michielin, O.: EADock: docking of small molecules into protein active sites with a multiobjective evolutionary optimization. Proteins **67**(4), 1010–1025 (2007)
6. Gu, J., Yang, X., Kang, L., Wu, J., Wang, X.: MoDock: a multi-objective strategy improves the accuracy for molecular docking. Algorithms Mol. Biol. **10**, 8 (2015)
7. Janson, S., Merkle, D., Middendorf, M.: Molecular docking with multi-objective particle swarm optimization. Appl. Soft Comput. **8**(1), 666–675 (2008)
8. Kukkonen, S., Lampinen, J.: GDE3: the third evolution step of generalized differential evolution. In: The 2005 IEEE Congress on Evolutionary Computation, vol. 1, pp. 443–450 (2005)
9. Li, H., Zhang, Q.: Multiobjective optimization problems with complicated pareto sets, MOEA/D and NSGA-II. IEEE Trans. Evol. Comput. **13**(2), 229–242 (2009)
10. López-Camacho, E., García-Godoy, M.J., Nebro, A.J., Aldana-Montes, J.F.: jMetalCpp: optimizing molecular docking problems with a C++ metaheuristic framework. Bioinformatics **30**(3), 437–438 (2014)
11. López-Camacho, E., García-Godoy, M.J., García-Nieto, J., Nebro, A.J., Aldana-Montes, J.F.: Solving molecular flexible docking problems with metaheuristics: a comparative study. Appl. Soft Comput. **28**, 379–393 (2015)
12. Morris, G.M., Huey, R., Lindstrom, W., Sanner, M.F., Belew, R.K., Goodsell, D.S., Olson, A.J.: AutoDock4 and AutoDockTools4: automated docking with selective receptor flexibility. J. Comput. Chem. **30**(16), 2785–2791 (2009)
13. Nebro, A.J., Durillo, J.J., Garcia-Nieto, J., Coello Coello, C.A., Luna, F., Alba, E.: SMPSO: a new PSO-based metaheuristic for multi-objective optimization. In: IEEE Symposium on Computational Intelligence in Multi-criteria Decision-Making, pp. 66–73, March 2009
14. Norgan, A.P., Coffman, P.K., Kocher, J.P.A., Katzmann, D.J., Sosa, C.P.: Multilevel parallelization of AutoDock 4.2. J. Cheminform. **3**(1), 12 (2011)
15. Oduguwa, A., Tiwari, A., Fiorentino, S., Roy, R.: Multi-objective optimisation of the protein-ligand docking problem in drug discovery. In: Proceedings of the 8th Annual Conference on Genetic and Evolutionary Computation, pp. 1793–1800 (2006)
16. Sandoval-Perez, A., Becerra, D., Vanegas, D., Restrepo-Montoya, D., Nino, F.: A multi-objective optimization energy approach to predict the ligand conformation in a docking process. In: Krawiec, K., Moraglio, A., Hu, T., Etaner-Uyar, A.Ş., Hu, B. (eds.) EuroGP 2013. LNCS, vol. 7831, pp. 181–192. Springer, Heidelberg (2013)

17. Sheskin, D.J.: Handbook of Parametric and Nonparametric Statistical Procedures. Chapman & Hall/CRC, New York (2007)
18. Zhang, Q., Li, H.: MOEA/D: a multiobjective evolutionary algorithm based on decomposition. IEEE Trans. Evol. Comput. **11**(6), 712–731 (2007)

Gibbs/MCMC Sampling for Multiple RNA Interaction with Sub-optimal Solutions

Saad Mneimneh$^{(\boxtimes)}$ and Syed Ali Ahmed

Hunter College and The Graduate Center City University of New York,
New York, NY, USA
saad@hunter.cuny.edu, sahmed3@gradcenter.cuny.edu

Abstract. The interaction of two RNA molecules involves a complex interplay between folding and binding that warranted the development of RNA-RNA interaction algorithms. However, these algorithms do not handle more than two RNAs. We note our recent successful formulation for the multiple (more than two) RNA interaction problem based on a combinatorial optimization called *Pegs and Rubber Bands*. Even then, however, the optimal solution obtained does not necessarily correspond to the actual biological structure. Moreover, a structure produced by interacting RNAs may not be unique to start with. Multiple solutions (thus sub-optimal ones) are needed. Here, a sampling approach that extends our previous formulation for multiple RNA interaction is developed. By clustering the sampled solutions, we are able to reveal representatives that correspond to realistic structures. Specifically, our results on the U2-U6 complex and its introns in the spliceosome of yeast, and the CopA-CopT complex in E. Coli are consistent with published biological structures.

Keywords: Multiple RNA interaction · RNA structure · Gibbs sampling · Metropolis-Hastings algorithm · Clustering

1 Introduction

The role of interaction between two or more RNA molecules has been increasingly recognized in regulatory mechanisms, including gene expression, methylation, and splicing. Pairwise interaction has been noted for regulating gene expression, e.g. when one RNA binds to the ribosome binding site of another mRNA, thus blocking its translation to protein [18]. Typical scenarios of multiple RNA interaction involve the interaction of multiple small nucleolar RNAs (snoRNAs) with ribosomal RNAs (rRNAs) in guiding the methylation of the rRNAs [24], and multiple small nuclear RNAs (snRNA) with mRNAs in the splicing of introns [34].

The prediction of structures resulting from pairwise interactions is now somewhat understood, due to successful efforts in generalizing the *partition function* of a single RNA to the case of two.

S. Mneimneh—Supported by a Research Starter Award in Informatics from the PhRMA Foundation. Partially supported by CoSSMO CUNY.

S.A. Ahmed—Supported by a PSC CUNY Award 68671-00 46.

© Springer International Publishing Switzerland 2016
M. Botón-Fernández et al. (Eds.): AlCoB 2016, LNBI 9702, pp. 78–90, 2016.
DOI: 10.1007/978-3-319-38827-4_7

Algorithms for pairwise interaction of RNAs can be found in [3, 7, 8, 15, 19, 24, 25, 29, 31, 32]. However, when carried over to multiple RNAs (more than two), generalizing the partition function further does not necessarily lead to efficient algorithms for computing it. Consequently, structure prediction in the context of multiple RNAs was almost non-existent; with just a few attempts that lack the ability to produce realistic structures. The de facto approach for multiple RNAs has been to account for their interaction by concatenating the RNAs into a single long RNA, which is then folded in order to predict the structure [4, 10]. On the one hand, this presents a challenge to existing folding algorithms, which are far less reliable when the RNA is too long. On the other hand, most folding algorithms prevent the formation of pseudoknots due to their increased computational complexity. While pseudoknots are rare in folded structures, they translate into kissing loops when spanning multiple RNAs, which are quite frequent in interacting RNA structures. There are a few attempts for introducing kissing loops into the concatenation model, e.g. [6], but advances in pairwise interaction algorithms based on the generalized partition function suggest that the latter are more adequate, so they remain the state-of-the-art for two RNAs.

Therefore, a promising approach is to adapt existing pairwise interaction algorithms to the case of multiple RNAs. This generally leads to a computational hurdle: when RNAs are treated pairwise, an immediate consequence is the *greedy* nature of the algorithm. The best interacting pair of RNAs will dominate the solution, as in [35, 36]. Since the pair of RNAs is required to fully interact, this will "lock" the interaction pattern of the whole ensemble into a sub-optimal state; thus preventing the correct structure from presenting itself as a solution.

We have recently proposed in a series of works [1, 2, 26, 28] a mathematical formulation based on combinatorial optimization that overcomes the issues outlined above. The model handles multiple RNAs without having to generalize the partition function beyond pairs. The resulting algorithms are not based on the concatenation paradigm, so they allow the formation of kissing loops, as well as other structures. And while they are still primarily based on an adaptation of pairwise interaction, they avoid the "locking" problem mentioned earlier.

Even then, obtaining one (optimal) solution for a multiple RNA interaction problem is not completely satisfactory. Many biological factors are hard to account for computationally. In addition, correct biological structures are often not unique. Therefore, some realistic solutions are ought to be sub-optimal, which is what we address here.

2   Preliminaries

2.1   The Model: Pegs and Rubber Bands

We advocate a combinatorial optimization problem called Pegs and Rubber Bands as a framework for multiple RNA interaction. The link between the two will be made shortly following a formal description of Pegs and Rubber Bands.

Consider m levels numbered 1 to m with n_l pegs in level l numbered 1 to n_l. There is an infinite supply of rubber bands, and a rubber band can be placed around pegs in consecutive levels. For instance, we may choose to place a rubber band around pegs $[i_1, i_2]$ (i.e., the set of pegs from i_1 to i_2, where $i_1 \leq i_2$), in level l, and pegs $[j_1, j_2]$ in level $l + 1$. In this case, the rubber band defines a window with a given weight $w(l, i_2, j_2, u, v)$, where $u = i_2 - i_1 + 1$ and $v = j_2 - j_1 + 1$ represent the lengths of the intervals covered by the window in levels l and $l + 1$, respectively (as in Fig. 1). For convenience, we will use $w(l, i, j, u, v)$ interchangeably to denote both the window and its weight, depending on context. As such, each window $w(l, i, j, u, v)$ defines two intervals, $[i - u + 1, i]$ in level l and $[j - v + 1, j]$ in level $l + 1$. Two windows overlap if any of their intervals overlap on the same level. In addition, $w(l, i, j, u, v)$ and $w(l, i', j', u', v')$ overlap if $\text{sgn}(i - i') \neq \text{sgn}(j - j')$ (their rubber bands cross).

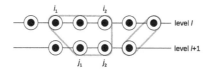

Fig. 1. A rubber band around pegs defines a window. The lengths $u = i_2 - i_1 + 1$ and $v = j_2 - j_1 + 1$ of the corresponding intervals may be different.

The Pegs and Rubber Bands problem is to maximize the total weight by placing rubber bands around pegs in such a way that none of their corresponding windows overlap.

To make the connection with multiple RNA interactions: RNA sequences become the levels, the ordered pegs in each level represent RNA bases $\{A, G, C, U\}$ in the order of occurrence in their sequence, a window $w(l, i, j, u, v)$ is an interaction between bases $[i - u + 1, i]$ in RNA l and bases $[j - v + 1, j]$ in RNA $l + 1$, and the weight $w(l, i, j, u, v)$ is chosen based on the energy of that interaction. The energies are obtained using a generalized partition function for pairwise interaction, and account for both intra- and inter- molecular energies. The no overlap condition reflects a typical nature of RNA interactions, and the maximization nature of the problem corresponds to energy minimization.

2.2   An Approximation Algorithm

A polynomial time approximation scheme (PTAS) for Pegs and Rubber Bands based on dynamic programming was described in [2, 28], where $n = \max_l n_l$.

Theorem 1. *Polynomial Time Approximation Scheme (PTAS) Pegs and Rubber Bands is NP-hard; however, for every $\epsilon > 0$, it admits a polynomial time algorithm that runs in $O(\lceil \frac{1}{\epsilon} \rceil mn^{\lceil \frac{1}{\epsilon} \rceil})$ time and achieves a total weight within a $(1 - \epsilon)$-factor of optimal.*

The mapping of RNAs to levels can be obtained as in [2, 28]. Figure 2 shows an example of a structure predicted using the Pegs and Rubber Bands formulation as reported in [2, 28], where windows are replaced by bonds between their

```
I1    3' UGUAUG 5'
         ||||
U6    5' AUAC...GAUU...GUGAAGCGU 3'
             ||||    |||||||||
U2    3' UAUGAU...CUAG...CACUUCGCA 5'
         |||||
I2    5' UACUAAC 3'
```

Fig. 2. Multiple RNA interaction within the eukaryotic spliceosome, a large ribonu-cleoprotein assembly responsible for the excision of intervening sequences in precursor messenger (pre-mRNA) molecules. Showing is the spliceosomal U2-U6 small nuclear (snRNA) and introns I1 and I2. The resulting structure is consistent with biological experiments [34,38].

corresponding intervals. The formulation avoids the "locking" problem, since treating the RNAs pairwise would have favored the full binding of U2-U6 to include their left extremities in Fig. 2, leaving I1 and I2 detached.

3   Realistic Biological Factors and Sub-optimal Solutions

Most algorithms for RNA-RNA interaction compute a partition function for the two RNAs based on loop energies in ways inspired by the basic algorithm of McCaskill for a single RNA [21]. Thus, when it comes to multiple RNA inter-action, the maximization of weight in the Pegs and Rubber Bands problem is somewhat equivalent to minimization of energy.

We have successfully used weights obtained from the tool RNAup [29] as follows: $w(l, i, j, u, v) \propto \log P_l(free[i - u + 1, i]) + \log P_{l+1}(free[j - v + 1, j]) + \log Z_l^I(i - u + 1, i, j - v + 1, j)$ where $P_l(free[i,j])$ is the probability that sub-sequence $[i, j]$ is free (does not fold) in RNA l, and $Z_l^I(i_1, i_2, j_1, j_2)$ is the gen-eralized partition function of the interaction of subsequences $[i_1, i_2]$ in RNA l and $[j_1, j_2]$ in RNA $l + 1$ (subject to no folding within the RNAs subsequences). Therefore, the method may be categorized as an MFE-like approach (Minimum Free Energy). It is clear that such an approach does not capture "everything".

Many biological factors affect the observed structure of interacting RNA molecules. For instance, reversible kissing loops (where some hydrogen bonds of the interaction between hairpins unwind) [17] are generally not captured by MFE since a kissing loop is energetically more favorable than a partial one. We observe such artifacts within the pairwise interaction of CopA-CopT in E. Coli, as shown in Fig. 3.

Another example is the U2-U6 snRNA complex. There seems to be a lack of consensus whether the U2-U6 snRNA complex forms a 4-way or a 3-way junction (most likely both structures co-exist [5,30,33,38]). Figure 4 shows the two possibilities. It has been conjectured in [5] that co-axial stacking is essential for the stabilization of helix I in U2-U6 and, therefore, inhibition of the co-axial stacking, possibly by protein binding, may activate the second conformation (with helices Ia and Ib).

(a)

```
CopA 5' CGGUUUAAGUGGG...UUUCGUACUCGCCAAAGUUGAAGA...UUUUGCUU 3'
        |||||||||||||   ||||||||||||||||||||||||   ||||||||
CopT 3' GCCAAAUUCACCC...AAAGCAUGAGCGGUUUCAACUUCU...AAAACGAA 5'
```

(b)

```
CopA 5' CGGUUUAAGUGGG...UUUCGUACUCGCCAAAGUUGAAGA...UUUUGCUU 3'
        |||||||||||||   |||||||||   ||||||
CopT 3' GCCAAAUUCACCC...AAAGCAUGAGCGGUUUCAACUUCU...AAAACGAA 5'
```

Fig. 3. The pairwise interaction of CopA-CopT: (a) computational prediction with artifact interactions due to the maximization nature of the problem, and (b) the actual biologically known interaction [18], where the last window is dropped and the middle window is split (reversible kissing loop).

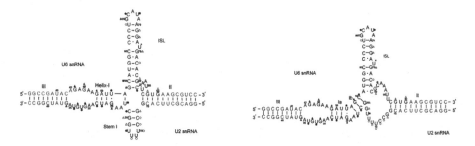

Fig. 4. U2-U6 snRNA complex in humans obtained by Greenbaum's lab [38]. The 4-way junction appears on the left hand side with Helix I, and the 3-way junction appears on the right hand side with Helices Ia and Ib.

Therefore, correct biological structures are not always "optimal" (from the computational perspective), and often are not unique. Sub-optimal solutions are needed to cover the biological ground. To that end, we consider in this paper two main modifications to our original approach based on Pegs and Rubber Bands:

– Sampling is used to produce multiple (sub-optimal) solutions instead of a single solution (this is described in Sect. 4).
– Windows are considered to be either *single* or *dependent*. Single windows contribute a weight equal to a sum of three terms as described above (our original formulation). Recall that each window $w(l, i, j, u, v)$ defines two intervals, $[i - u + 1, i]$ in level l and $[j - v + 1, j]$ in level $l + 1$. If a solution contains two windows that define intervals $[a, b]$ and $[c, d]$ in level l with $b < c$ and no other intervals in between, then we may consider them dependent in level l (windows can be dependent in one or two levels) and thus replace $\log P_l(free[a, b]) + \log P_l(free[c, d])$ (each of these two terms is coming from the *single* contribution of each window) with $\log P_l(free[a, d])$, if the latter is larger than the former sum. This allows window splits such as the one shown

in Fig. 3(b) to be not so detrimental to the total weight of the solution. Given a solution, its total weight is then obtained by the optimal determination of single and dependent windows in each level to maximize that weight (this is achieved by a dynamic programming algorithm for each level). We denote this modified weight of a solution S by $w(S)$.

4   A Sampling Approach

Sampling is more efficient than exhaustive enumeration of solutions within a certain threshold of optimal, especially that many of these solutions will be similar. Furthermore, sampling has been successfully used in the context of a single RNA; for instance, in [9,23,37] to mention a few examples. For the multiple RNA interaction, we propose below an approach based on Gibbs sampling and the Metropolis-Hastings algorithm.

4.1   The Gibbs Sampler

The described model for multiple RNA interaction, viewed as Pegs and Rubber Bands with m levels, lends itself quite naturally to Gibbs sampling [13,20]. As a random variable, let S_l be a set of non-overlapping windows of the form $w(l, i, j, u, v)$, so S_l represents a valid interaction pattern between RNA l and RNA $l+1$. A Gibbs sampler works by sampling each random variable individually in order, conditioned on the current values of the other variables. In other words, we work with $P(S_l|S_1, \ldots, S_{l-1}, S_{l+1}, \ldots, S_{m-1})$. Therefore, if we start with $S_1^0 = \ldots = S_{m-1}^0 = \emptyset$, we sample S_1^1 using $P(S_1|S_2^0, \ldots, S_{m-1}^0)$, then S_2^1 using $P(S_2|S_1^1, S_3^0, \ldots S_{m-1}^0)$, then S_3^1 using $P(S_3|S_1^1, S_2^1, S_4^0, \ldots, S_{m-1}^0)$, and so on until we sample S_{m-1}^1 using $P(S_{m-1}|S_1^1, \ldots, S_{m-2}^1)$. We call $(S_1^1, \ldots, S_{m-1}^1)$ our first sample, and we repeat to obtain $(S_1^t, \ldots, S_{m-1}^t)$ for every t. Under typical conditions of ergodicity [11], the Gibbs guarantee is that $(S_1^t, \ldots, S_{m-1}^t)$ for large t is a sample from $P(S_1, \ldots, S_{m-1})$, which is not necessarily a known distribution, in contrast to $P(S_l|S_1, \ldots, S_{l-1}, S_{l+1}, \ldots, S_{m-1})$ which is reasonably accessible.

This is interesting because, conditioned on $S_1, \ldots, S_{l-1}, S_{l+1}, \ldots, S_{m-1}$, the permissible windows of the form $w(l, i, j, u, v)$ are exactly those which do not overlap with windows in S_{l-1} and S_{l+1}. As such, we assume that:

$$P(S_l|S_1, \ldots, S_{l-1}, S_{l+1}, \ldots, S_{m-1}) = P(S_l|S_{l-1}, S_{l+1})$$

$$P(S_l|S_{l-1}, S_{l+1}) \propto \begin{cases} 0 & S_l \text{ contains a window that overlaps in } S_{l-1} \text{ or } S_{l+1} \\ e^{w(S_l)} & \text{otherwise} \end{cases}$$

The exponential term is similar in spirit to the standard Boltzman distribution used for RNAs, knowing that $w(S_l)$ represents the negative of the energy.

If $P(S_l|S_{l-1}, S_{l+1})$ is easy to sample from, then the Gibbs sampler works nicely given a fixed mapping of RNAs to levels 1 to m. We describe in the next section how to sample from $P(S_l|S_{l-1}, S_{l+1})$.

4.2   Gibbs Sampling with Metropolis-Hastings

The Metropolis-Hastings algorithm for sampling (also known as the Markov Chain Monte Carlo method) was described in [14, 22], and since then has been utilized extensively in the literature. To sample from $P(S_l|S_{l-1}, S_{l+1})$, we first drop all the windows of the form $w(l, i, j, u, v)$ that overlap in S_{l-1} or S_{l+1}. We only work with the remaining windows of the form $w(l, i, j, u, v)$. We then construct a random sequence $S_l^0, S_l^1, \ldots$, where S_l^t is a set of non-overlapping windows of the form $w(l, i, j, u, v)$. This can be done with a Metropolis-Hastings strategy: Given S_l^t, we randomly generate S_l^{t+1} with some proposal probability $Q(S_l^{t+1}|S_l^t)$, and either accept S_l^{t+1} with probability

$$\min\left\{1, \frac{Q(S_l^t|S_l^{t+1})}{Q(S_l^{t+1}|S_l^t)} \times \frac{e^{w(S_l^{t+1})}}{e^{w(S_l^t)}}\right\}$$

or reject it and let $S_l^{t+1} = S_l^t$.

It is well known and easy to show that such a strategy results in a Markov chain which converges to the desired probability distribution if the proposal chain $Q(S_l^{t+1}|S_l^t)$ satisfies $Q(S_l^{t+1} = y|S_l^t = x) > 0 \Leftrightarrow Q(S_l^{t+1} = x|S_l^t = y) > 0$; this also makes it irreducible [12].

For practical purposes, we limit S_l^t to contain only windows $w(l, i, j, u, v)$ where $u = v$. We also do not allow two adjacent windows $w(l, i, j, u, v)$ and $w(l, i - u, j - v, u', v')$ to co-exists (since together they represent one bigger window). With that in mind, a simple strategy is to make $Q(S_l^{t+1}|S_l^t)$ **uniform** among all the neighbors of S_l^t (including S_l^t itself), where a neighbor other than S_l^t can be obtained by one of the following three operations:

- a window $w(l, i, j, u, v) \in S_l^t$ is removed from S_l^t
- a window $w(l, i, j, u, v) \notin S_l^t$ that does not overlap in S_l^t is added to S_l^t
- a window $w(l, i, j, u, v) \in S_l^t$ is replaced by a window $w(l, i', j', u', v') \notin S_l^t$ that only overlaps with $w(l, i, j, u, v)$ in S_l^t

Therefore, for every S_l^{t+1} that is a neighbor of S_l^t, $Q(S_l^{t+1}|S_l^t)$ is the inverse of the number of neighbors of S_l^t. This proposal probability defines an irreducible Markov chain since every pair of solutions can be reached from one another through a sequence of neighbors.

4.3   A Notion of Distance for Sub-optimal Solutions

Many of the sampled sub-optimal solutions will be similar. To quantify this similarity/dissimilarity, we need to describe a distance function. To motivate our approach, we first define the notion of a *terminal* window: Given a solution S, the terminal window $w(l, i, j, u, v) \in S$ is the window with the largest l such that no windows appear on its right in levels $l - 1$, l, and $l + 1$:

- no window $w(l - 1, i', j', u', v') \in S$ has $j' > i$
- no window $w(l, i', j', u', v') \in S$ has $i' > i$
- no window $w(l + 1, i', j', u', v') \in S$ has $i' > j$

By recursively eliminating the terminal window from a solution, we obtain a total order on the windows of that solution.

Our approach builds on the idea that if two solutions are similar, we expect them to have a similar set of windows; furthermore, these windows should exhibit the same order. In more detail, given a solution S, define $|S|$ as the number of windows in S, and let $w(l_1, i_1, j_1, u_1, v_1), \ldots, w(l_{|S|}, i_{|S|}, j_{|S|}, u_{|S|}, v_{|S|})$ be the $|S|$ windows in the order defined by terminal windows. Each of these windows, say $w(l, i, j, u, v)$, defines the two intervals, $[i - u + 1, i]$ in level l and $[j - v + 1, j]$ in level $l + 1$. Define the set of interaction intervals

$$I(S) = (I_1, \ldots, I_{2|S|}) = ([i_1 - u_1 + 1, i_1], [j_1 - v_1 + 1, j_1], \ldots$$

$$\ldots, [i_{|S|} - u_{|S|} + 1, i_{|S|}], [j_{|S|} - v_{|S|} + 1, j_{|S|}])$$

as an ordered sequence of $2|S|$ intervals, and $L(S) = (l_1, \ldots, l_{|S|})$ as an ordered sequence of $|S|$ levels, where l_i is the level defining the ith window. Therefore, $L(S)$ means that we have the following set of pairwise interactions (not necessarily unique in terms of RNAs): RNA l_1 with RNA $l_1 + 1$, RNA l_2 with RNA $l_2 + 1$, ..., RNA $l_{|S|}$ with RNA $l_{|S|} + 1$. Two solutions that do not agree on this set, or do not define overlapping interaction intervals, are considered completely dissimilar; otherwise, their distance is given by the amount of overlap in their interaction intervals (as in the Jaccard metric [16]), hence the following definition of distance:

Given two solutions S_1 with $I(S_1) = (I_1, I_2, \ldots)$ and S_2 with $I(S_2) = (T_1, T_2, \ldots)$, the distance between S_1 and S_2 is

$$d(S_1, S_2) = \begin{cases} 1 - \frac{\sum_i |I_i \cap T_i|}{\sum_i |I_i \cup T_i|} & L(S_1) = L(S_2) \text{ and } I_i \cap T_i \neq \emptyset \text{ for all } i \\ 1 & \text{otherwise} \end{cases}$$

where $\cap$ and $\cup$ represent the standard intersection and union operations on sets respectively, and intervals are treated as sets of integers. This distance is modified from our previous metric in [26,27], and is not a metric; however, it works well with the clustering algorithm described below.

4.4   Clustering the Samples

The sampled sub-optimal solutions are generally more than what we need. In addition, as mentioned above, many of them will be similar. Therefore, we use clustering to reduce their number. To cluster the samples, we first remove duplicates, so we only work with unique samples. We then drop all solutions with a weight below 1/3 of the best. Finally, we sort the solutions to make the output of the clustering deterministic. We adopt hierarchical agglomerative clustering with complete linkage, and we obtain the clusters by "cutting" the tree where distance between clusters is 1. Given the clusters, the optimal solution in each cluster acts as a "representative" of the cluster. The representatives should reveal some of the structures that are observed in biological experiments [1,26,27].

5   Experimental Results

We perform 50 iterations of the Metropolis-Hastings algorithm **without** rejection. This allows us to start at some random solution. We then allow 50 iterations (with rejection) for the "burn-in" time of the Metropolis-Hastings algorithm. Finally, we generate 50 samples in 50 iterations and select one uniformly at random. We generate 1000 solutions (Gibbs samples) by repeating this procedure, as described in Sect. 4.1.

After clustering, we sort the representatives of the clusters by decreasing weight. We consider the first k representatives, for a given k. To assess our approach, we repeat the experiment 100 times. Given a set of candidate structures in mind; for instance, Fig. 5 shows four candidates for the yeast spliceosome, we then count how many times (in the 100 runs) each candidate is found among the first k representatives, as a percentage hit. We also compute the "rank" of each candidate, which is the first time[1] that candidate is seen as representative, averaged over the 100 experiments.

5.1   Experiment 1: Structural Variation

The U2-U6 complex in the spliceosome of yeast has been reported to have two distinct experimental structures, e.g. [33]. In one conformation, U2 and U6 interact to form a helix known as helix Ia. In another conformation, the interaction reveals a structure containing an additional helix, known as helix Ib. Section 3 describes possible underlying mechanisms that are responsible for this conformational switch. We consider the set of four candidates in Fig. 5. The results are summarized in Table 1.

Table 1. Results for the yeast spliceosome. Each entry lists the percentage hit followed by the average rank.

k	1		2		3		4		5		6		7		8		9		10	
Helices Ia+Ib	100	1	100	1	100	1	100	1	100	1	100	1	100	1	100	1	100	1	100	1
Helices Ia+Ib, I1 detached	0	–	100	2	100	2	100	2	100	2	100	2	100	2	100	2	100	2	100	2
Helix Ia	0	–	0	–	0	–	40	4	85	4.5	100	4.8	100	4.8	100	4.8	100	4.8	100	4.8
Helix Ia, I1 detached	0	–	0	–	0	–	0	–	40	5	85	5.5	100	5.8	100	5.8	100	5.8	100	5.8

5.2   Experiment 2: Artifact Interactions

Due to the optimization nature of the problem, it is sometimes easy to pick up interactions that are not biologically real. This is because dropping these interactions from the solution would make it sub-optimal (even when preferred biologically, as described in Sect. 3). The last interaction window of CopA-CopT

---

[1] We use "first time" because many solutions can represent the same candidate; for instance, a window can split in different ways, but we still refer to it as a window split.

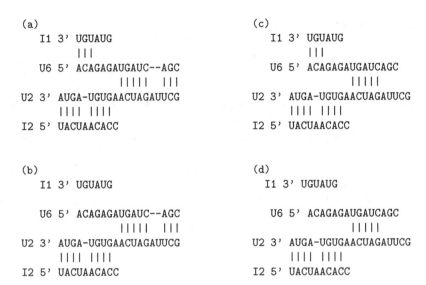

Fig. 5. The yeast spliceosome with 4 RNAs (I1 and I2 are functionally independent stretches of the same much longer messenger RNA). (a) Helix Ia and helix Ib with both introns attached. (b) Helix Ia and helix Ib with I1 detached. (c) Helix Ia with both introns attached. (d) Helix Ia with I1 detached.

in Fig. 3 is an example of such an artifact. We consider six candidate solutions based on presence/absence of windows and window splits, as described in Table 2. For each of the three interaction windows in Fig. 3, we consider whether the window is present, dropped, or split. Typically, we detect a window split when the two portions happen to be treated as *dependent* in some level l (see Sect. 3). Therefore, to correctly capture reversible kissing loops, undesired splits can be ignored if the corresponding window does not represent a kissing loop. Given the RNA structures of CopA and CopT, only the middle window is a kissing loop.

Table 2. Results for CopA-CopT. For each of the three interaction windows in Fig. 3, we consider whether the window is present, dropped, or split. Each entry lists the percentage hit followed by the average rank.

k	1		2		3		4		5		6		7		8		9		10	
First, middle, last	89.6	1	93.8	1	97.9	1.1	100	1.2	100	1.2	100	1.2	100	1.2	100	1.2	100	1.2	100	1.2
First, middle split, last	4.2	1	52.1	1.9	77.1	2.3	93.8	2.6	97.9	2.7	100	2.8	100	2.8	100	2.8	100	2.8	100	2.8
First, middle, last dropped	4.2	1	10.4	1.6	16.7	2.1	20.8	2.5	27.1	3.1	29.2	3.3	31.2	3.5	31.2	3.5	35.4	4.2	35.4	4.2
First, middle split, last dropped	0	–	2.1	2	2.1	2	4.2	3	12.5	4.3	18.8	4.9	25	5.4	29.2	5.8	37.5	6.5	41.7	6.8
First split, middle, last	2.1	1	8.3	1.8	20.8	2.5	27.1	2.8	43.8	3.7	54.2	4.1	70.8	4.8	79.2	5.1	83.3	5.3	83.3	5.3
First split, middle, last dropped	0	–	0	–	0	–	0	–	2.1	5	4.2	5.5	6.2	6	6.2	6	10.4	7.2	10.4	7.2

6   Conclusion

In RNA interaction, the optimal structure may not be the real structure, and the real structure may not be unique. In this work, we build on our previous approach for multiple RNA interaction using the Pegs and Rubber Bands formulation to generate multiple sub-optimal solutions. This is developed using Gibbs sampling and the Metropolis-Hastings algorithm.

Our sampling approach successfully computes sub-optimal solutions for the multiple RNA interaction problem that are truthful representations of the actual biological structures. For instance, it can provide several candidate structures when they exist, e.g. the U2-U6 complex and its introns in the spliceosome of yeast, and find structures that agree with the literature, but are not necessarily optimal in the computational sense, e.g. CopA-CopT in E. Coli.

References

1. Ahmed, S.A., Mneimneh, S.: Multiple RNA interaction with sub-optimal solutions. In: Basu, M., Pan, Y., Wang, J. (eds.) ISBRA 2014. LNCS, vol. 8492, pp. 149–162. Springer, Heidelberg (2014)
2. Ahmed, S.A., Mneimneh, S., Greenbaum, N.L.: A combinatorial approach for multiple RNA interaction: formulations, approximations, and heuristics. In: Du, D.-Z., Zhang, G. (eds.) COCOON 2013. LNCS, vol. 7936, pp. 421–433. Springer, Heidelberg (2013)
3. Alkan, C., Karakoc, E., Nadeau, J.H., Sahinalp, S.C., Zhang, K.: RNA-RNA interaction prediction and antisense RNA target search. J. Comput. Biol. **13**(2), 267–282 (2006)
4. Andronescu, M., Zhang, Z.C., Condon, A.: Secondary structure prediction of interacting RNA molecules. J. Mol. Biol. **345**(5), 987–1001 (2005)
5. Cao, S., Chen, S.J.: Free energy landscapes of RNA/RNA complexes: with applications to snRNA complexes in spliceosomes. J. Mol. Biol. **357**(1), 292–312 (2006)
6. Chen, H.L., Condon, A., Jabbari, H.: An $o(n^5)$ algorithm for MFE prediction of kissing hairpins and 4-chains in nucleic acids. J. Comput. Biol. **16**(6), 803–815 (2009)
7. Chitsaz, H., Backofen, R., Sahinalp, S.C.: biRNA: fast RNA-RNA binding sites prediction. In: Salzberg, S.L., Warnow, T. (eds.) WABI 2009. LNCS, vol. 5724, pp. 25–36. Springer, Heidelberg (2009)
8. Chitsaz, H., Salari, R., Sahinalp, S.C., Backofen, R.: A partition function algorithm for interacting nucleic acid strands. Bioinformatics **25**(12), i365–i373 (2009)
9. Ding, Y., Lawrence, C.E.: A statistical sampling algorithm for RNA secondary structure prediction. Nucleic Acids Res. **31**(24), 7280–7301 (2003)
10. Dirks, R.M., Bois, J.S., Schaeffer, J.M., Winfree, E., Pierce, N.A.: Thermodynamic analysis of interacting nucleic acid strands. SIAM Rev. **49**(1), 65–88 (2007)
11. Durbin, R., Eddy, S.R., Krogh, A., Mitchison, G.: Biological Sequence Analysis: Probabilistic Models of Proteins and Nucleic Acids. Cambridge University Press, Cambridge (1998). Chap. 11
12. Gallager, R.G.: Discrete Stochastic Processes, vol. 321. Springer Science & Business Media, Newyork (2012). Chap. 4

13. Geman, S., Geman, D.: Stochastic relaxation, gibbs distributions, and the bayesian restoration of images. IEEE Trans. Pattern Anal. Mach. Intell. PAMI **6**(6), 721–741 (1984)
14. Hastings, W.K.: Monte carlo sampling methods using markov chains and their applications. Biometrika **57**(1), 97–109 (1970)
15. Huang, F.W., Qin, J., Reidys, C.M., Stadler, P.F.: Partition function and base pairing probabilities for RNA-RNA interaction prediction. Bioinformatics **25**(20), 2646–2654 (2009)
16. Jaccard, P.: Etude comparative de la distribution florale dans une portion des Alpes et du Jura. Impr. Corbaz (1901)
17. Kolb, F.A., Engdahl, H.M., Slagter-Jäger, J.G., Ehresmann, B., Ehresmann, C., Westhof, E., Wagner, E.G.H., Romby, P.: Progression of a loop-loop complex to a four-way junction is crucial for the activity of a regulatory antisense RNA. EMBO J. **19**(21), 5905–5915 (2000)
18. Kolb, F.A., Malmgren, C., Westhof, E., Ehresmann, C., Ehresmann, B., Wagner, E., Romby, P.: An unusual structure formed by antisense-target RNA binding involves an extended kissing complex with a four-way junction and a side-by-side helical alignment. RNA **6**(3), 311–324 (2000)
19. Li, A.X., Marz, M., Qin, J., Reidys, C.M.: RNA-RNA interaction prediction based on multiple sequence alignments. Bioinformatics **27**(4), 456–463 (2011)
20. Liu, J.S.: The collapsed gibbs sampler in bayesian computations with applications to a gene regulation problem. J. Am. Stat. Assoc. **89**(427), 958–966 (1994)
21. McCaskill, J.S.: The equilibrium partition function and base pair binding probabilities for RNA secondary structure. Biopolymers **29**(6–7), 1105–1119 (1990)
22. Metropolis, N., Rosenbluth, A.W., Rosenbluth, M.N., Teller, A.H., Teller, E.: Equation of state calculations by fast computing machines. J. Chem. Phys. **21**(6), 1087–1092 (1953)
23. Metzler, D., Nebel, M.E.: Predicting RNA secondary structures with pseudoknots by MCMC sampling. J. Math. Biol. **56**(1–2), 161–181 (2008)
24. Meyer, I.M.: Predicting novel RNA-RNA interactions. Curr. Opin. Struct. Biol. **18**(3), 387–393 (2008)
25. Mneimneh, S.: On the approximation of optimal structures for RNA-RNA interaction. IEEE/ACM Trans. Comput. Biol. Bioinf. (TCBB) **6**(4), 682–688 (2009)
26. Mneimneh, S., Ahmed, S.A.: Multiple RNA interaction: beyond two. To appear in IEEE Trans. Nanobiosci. (2015)
27. Mneimneh, S., Ahmed, S.A.: A sampling approach for multiple RNA interaction: finding sub-optimal solutions fast. In: BIOINFORMATICS 2015 - Proceedings of the International Conference on Bioinformatics Models, Methods and Algorithms, Rome, Italy, 21–23 February 2015
28. Mneimneh, S., Ahmed, S.A., Greenbaum, N.L.: Multiple RNA interaction - formulations, approximations, and heuristics. In: BIOINFORMATICS 2013 - Proceedings of the International Conference on Bioinformatics Models, Methods and Algorithms, Barcelona, Spain, 11–14 February 2013, pp. 242–249 (2013)
29. Mückstein, U., Tafer, H., Hackermüller, J., Bernhart, S.H., Stadler, P.F., Hofacker, I.L.: Thermodynamics of RNA-RNA binding. Bioinformatics **22**(10), 1177–1182 (2006)
30. Newby, M.I., Greenbaum, N.L.: A conserved pseudouridine modification in eukaryotic U2 snRNA induces a change in branch-site architecture. RNA **7**(6), 833–845 (2001)
31. Pervouchine, D.D.: IRIS: intermolecular RNA interaction search. Genome Inform. Ser. **15**(2), 92 (2004)

32. Salari, R., Backofen, R., Sahinalp, S.C.: Fast prediction of RNA-RNA interaction. Algorithms Mol. Biol. **5**(5) (2010)

33. Sashital, D.G., Cornilescu, G., Butcher, S.E.: U2–U6 RNA folding reveals a group II intron-like domain and a four-helix junction. Nat. Struct. Mol. Biol. **11**(12), 1237–1242 (2004)

34. Sun, J.S., Manley, J.L.: A novel U2–U6 snRNA structure is necessary for mammalian mRNA splicing. Genes Dev. **9**(7), 843–854 (1995)

35. Tong, W., Goebel, R., Liu, T., Lin, G.: Approximation algorithms for the maximum multiple RNA interaction problem. In: Widmayer, P., Xu, Y., Zhu, B. (eds.) COCOA 2013. LNCS, vol. 8287, pp. 49–59. Springer, Heidelberg (2013)

36. Tong, W., Goebel, R., Liu, T., Lin, G.: Approximating the maximum multiple RNA interaction problem. Theoret. Comput. Sci. **556**, 63–70 (2014)

37. Wei, D., Alpert, L.V., Lawrence, C.E.: Rnag: a new gibbs sampler for predicting RNA secondary structure for unaligned sequences. Bioinformatics **27**(18), 2486–2493 (2011)

38. Zhao, C., Bachu, R., Popović, M., Devany, M., Brenowitz, M., Schlatterer, J.C., Greenbaum, N.L.: Conformational heterogeneity of the protein-free human spliceosomal U2–U6 snRNA complex. RNA **19**(4), 561–573 (2013)

Phylogenetics

Accumulated Coalescence Rank and Excess Gene Count for Species Tree Inference

Sourya Bhattacharyya[✉] and Jayanta Mukhopadhyay

Department of Computer Science and Engineering,
Indian Institute of Technology, Kharagpur 721302, WB, India
sourya.bhatta@gmail.com, jay@cse.iitkgp.ernet.in

Abstract. We propose a novel summary based method to infer species trees from input multi-locus gene trees with incomplete lineage sorting (ILS). The method extends an existing technique called STAR [13], which defines average coalescence rank between taxa pairs (couplets), to derive species trees using Neighbor-Joining (NJ) [20,23]. Such coalescence rank, however, is ambiguous at couplet level. We propose two new couplet based distance measures, termed as *accumulated coalescence rank* (AcR), and *excess gene tree leaves* (XL), and show that their combination discriminates individual couplets better. We propose a new method **AcRN-JXL**, which uses the proposed measures, for NJ based species tree construction. Results show that for biological datasets, AcRNJXL produces much better performance than STAR and other reference approaches, with the same time and space complexities as STAR.

Keywords: Inferring the evolutionary phylogeny of species · Phylogeny reconstruction

1 Introduction

A gene tree depicts the evolutionary relationship of a given sample of gene copies obtained from a group of N taxa [5]. A set of M (>1) gene trees produced by sampling respective genes from N taxa, may be complete or partial, and may exhibit conflicting evolutionary histories (thus having conflicting topologies and branch lengths). Such conflict arises mainly due to one of the following three biological processes: (1) Horizontal gene transfer (HGT), (2) Gene duplication / loss, and (3) Incomplete Lineage Sorting (ILS) or Deep Coalescence (DC) [18]. Estimation of a species tree by modeling such gene tree / species tree discordance is essential to understand the evolutionary histories among these N taxa. Here we focus on the discordance caused by ILS, which occurs due to rapid speciation and short branches in respective gene trees, resulting failure of two or more lineages in a population to coalesce [5].

During ILS, most probable gene tree topology may not be the species tree, a condition termed as the 'anomaly zone' [5,18]. So, supertree or consensus approaches estimate comparatively less accurate species trees [28]. Traditional

© Springer International Publishing Switzerland 2016
M. Botón-Fernández et al. (Eds.): AlCoB 2016, LNBI 9702, pp. 93–105, 2016.
DOI: 10.1007/978-3-319-38827-4_8

concatenation (supermatrix) approach [10] employs phylogenetic reconstruction methods such as maximum likelihood (tools like RAxML [22]) to generate species trees from concatenated gene sequence alignments. But, concatenation may produce variable performances across datasets with high degree of ILS [14], and is also statistically inconsistent [19]. Statistical methods modeling multi-species coalescence, such as BEST [9], *BEAST [6], co-estimate gene and species trees from input gene alignments. Few other statistical methods like STEM [7], MP-EST [12], BUCKy [8], estimate species trees from gene trees, using maximum likelihood or Bayesian techniques. However, apart from MP-EST and STEM, these methods are computationally intensive [4], thus applicable to datasets involving ≈ 20 taxa and a few (< 50) gene trees [14].

Parsimony approaches aim to minimize either the sum of deep coalescence (MDC criterion) [1,25,28], or the sum of Robinson-Foulds (RF) distance between S and $\mathbf{G}$ (as in mulRF [2]). However, performance of these methods can vary for different degrees of ILS. Recently proposed summary methods ASTRAL [15] and ASTRAL-II [16] employ quartet decomposition and bi-partitions of input gene trees, to generate statistically consistent and highly accurate species trees. However, they incur high runtime for thousands of gene trees. Approaches GLASS [17], STEAC [13], employ couplet (taxa pair) based coalescence time for species tree inference. But such coalescence time information is not always accurate (or even not available) for many datasets [4]. Methods like STAR [13] and NJst [11] employ couplet based *coalescence rank* and *internode count* measures, respectively, for species tree inference. These methods are computationally efficient. But their performances on biological datasets need to be verified.

Current manuscript proposes an extension of STAR [13], by using two novel couplet based measures termed as *accumulated coalescence rank* (AcR), and *excess gene leaf count* (XL). We show that a combination of these measures discriminates individual couplets better. Subsequently, we describe a new method **AcRNJXL** using both of these measures for species tree inference. AcRNJXL supports incomplete, non-binary, multi-copy gene trees having ILS. It requires rooted gene trees to generate a rooted species tree. We show that AcRNJXL shows superior performance on biological datasets, with the same time and space complexities as STAR. Hence it is applicable for large scale biological datasets.

2   Methodology

2.1   Concept of Coalescence Rank and Basics of STAR

For a rooted gene tree G in the input gene tree set $\mathbf{G}$, suppose $L(G)$ and $NL(G)$ denote the set of leaves (taxa set) and the set of internal (non-leaf) nodes, respectively. For an internal node $n \in NL(G)$, Liu et al. [13] defined its *rank of a coalescence event, or simply the rank* $\lambda_G(n)$, as follows:

$$\lambda_G(n) = \begin{cases} |L(G)| & : n \text{ is the root node } r \\ |L(G)| - I(n, r) & : n \text{ is not the root node} \end{cases}$$

where $|.|$ is the set cardinality, and $I(n, r)$ is the number of internal nodes (including n) along the path between n and the root r. For a pair of taxa (x, y) (denoted as *couplets* from now on) covered in **G**, its *support set* τ_{xy} is defined as follows:

$$\tau_{xy} = \{G : (x, y) \in L(G)\} \tag{1}$$

For a couplet (x, y) in a gene tree G ($G \in \tau_{xy}$), let LCA_{xy}^G be their lowest common ancestor (LCA) in G. Then the *couplet coalescence rank* $R_G(x, y)$ for (x, y) in G is defined as the rank of LCA_{xy}^G ($= \lambda_G(LCA_{xy}^G)$) [13]. For a pair of couplets (x, y) and $(x, z) \in L(G)$, $R_G(x, y) < R_G(x, z)$ indicates that x and y coalesce before z. The *average couplet coalescence rank* $R_{avg}(x, y)$ for (x, y) is defined by the following equation [13]:

$$R_{avg}(x, y) = \frac{1}{|\tau_{xy}|} \sum_{G \in \tau_{xy}} R_G(x, y) \tag{2}$$

The method STAR [13] constructs a distance matrix D whose individual elements $d(x, y)$ (for a couplet (x, y)) are set to $R_{avg}(x, y)$. Subsequently, neighbor-joining (NJ) [20,23] is applied on D, to generate the species tree S.

2.2   Accumulated Coalescence Rank and Excess Gene Leaf Measures

Major drawback of the couplet coalescence rank is its ambiguity. For two pairs of couplets (x_1, y_1) and (x_2, y_2) in a gene tree G, $R_G(x_1, y_1) = R_G(x_2, y_2)$ if $LCA_{x_1 y_1}^G = LCA_{x_2 y_2}^G$. In general, if $|L(G)| = N$, $2 \leq R_G(x, y) \leq N$. Thus, mapping between couplets and their coalescence rank values has *cardinality ratio* of $\approx \frac{N^2}{2} : N = N : 2$. In view of this, we propose a new measure *accumulated couplet coalescence rank* (AcR) between individual couplets (x, y) of a gene tree G. This measure is denoted by $R'_G(x, y)$, and defined as following:

$$R'_G(x, y) = \sum_n \lambda_G(n) \tag{3}$$

where n denotes any internal node lying between the path from x to y via LCA_{xy}^G. Coalescence rank and the proposed AcR values for few couplets of the example phylogenetic tree (Fig. 1) are shown in Table 1.

For a gene tree G covering N taxa, values of $R'_G(x, y)$ for a couplet (x, y) exhibit following properties, based on whether G is a caterpillar or not (a caterpillar is a binary phylogenetic tree that reduces to a path once the leaf nodes and leaf edges are deleted).

1. $R'_G(x, y) \geq 2$.
2. $R'_G(x, y) \leq (\frac{N(N+1)}{2} - 1)$ (if G is a caterpillar).
3. $R'_G(x, y) \leq (2 \times \{ \sum_{N+1-\lceil \lg N \rceil}^{N-1} i \} + N)$ (for non-caterpillar G).

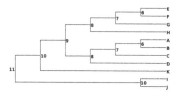

Fig. 1. An example phylogenetic tree. Labels associated with internal nodes denote their coalescence rank (λ) values.

Table 1. Values of coalescence rank [13], accumulated coalescence rank, and excess gene leaf count measures for some of the couplets of the tree.

Taxa pair	Coalescence rank (R)	Accumulated coalescence rank (R')	Excess gene leaf (X)
A,E	9	51	6
D,G	9	32	6
K,I	11	31	9
C,H	9	32	6

Proofs of the above properties are shown in Lemma 1 to 5 of the supplementary material (available in http://www.facweb.iitkgp.ernet.in/~jay/phtree/ AcRNJXL/AcRNJXL.html). Thus, mapping between the number of couplets and the proposed accumulated rank (R') has a cardinality ratio of $\approx N^2:N^2 = 1{:}1$. As R' has almost N times higher range of distribution than R, it discriminates individual couplets better. In general, couplets with higher AcR values are considered to be more distant, thus expected to coalesce later (closer to the root) compared to other couplets having lower AcR values. However, this observation does not always hold. For example, considering Fig. 1, the couplet (K,I) has lower AcR compared to the couplet (D,G), although the LCA of (K,I) is at higher level compared to the LCA of (D,G). Thus, sole use of the AcR values may not reveal the correct order of evolution.

Considering above drawback of AcR, we introduce another couplet based measure termed as the *number of excess gene tree leaves*, to use in conjunction with AcR. For a gene tree G, suppose a subtree rooted at an internal node $v \in NL(G)$ is denoted as $Clade_G(v)$. Further, let the set of taxa underlying $Clade_G(v)$ be represented as $Cluster_G(v)$. Then, the *number of excess leaves* $X_G(x,y)$ for a couplet (x,y) in G, is defined as follows:

$$X_G(x,y) = |Cluster_G(LCA_{xy}^G)| - 2 \tag{4}$$

For any couplet $(x,y) \in L(G)$, $0 \le X_G(x,y) \le (N-2)$. Lower $X_G(x,y)$ means that (x,y) coalesces earlier in G, compared to other couplets having higher excess leaf counts. Although XL measure is an indicator of the order of evolution among different couplets, it is also ambiguous. Couplets with identical LCA node have the same XL value. So, we propose to use both AcR and XL measures for couplet based species tree estimation. Couplets having high values of both AcR and XL measures are distant. Couplets having low values of both AcR and XL measures are evolutionarily very close.

For a particular couplet (x,y), *average excess gene leaf count* $X_{avg}(x,y)$ with respect to the input gene tree set $\mathbf{G}$, is defined as following:

$$X_{avg}(x,y) = \frac{1}{|\tau_{xy}|} \sum_{G \in \tau_{xy}} X_G(x,y) \tag{5}$$

Lower $X_{avg}(x, y)$ means (x, y), on the average, coalesces early in $\mathbf{G}$, than other couplets having higher average excess gene leaf counts. Suppose, N is the number of taxa covered in $\mathbf{G}$. Then, we construct an $N \times N$ distance matrix D_X, whose individual elements $d_X(x, y)$ are set to $X_{avg}(x, y)$.

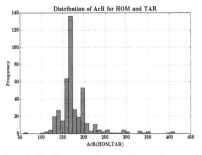
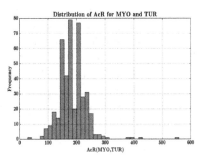

(a) AcR distribution of (HOM, TAR)   (b) AcR distribution of (MYO, TUR)

Fig. 2. Example of accumulated coalescence rank (R') distribution between two different couplets of the Mammalian dataset [15, 21].

Similarly, the *average accumulated coalescence rank* $R'_{avg}(x, y)$ for a couplet (x, y) is defined as the following:

$$R'_{avg}(x, y) = \frac{1}{|\tau_{xy}|} \sum_{G \in \tau_{xy}} R'_G(x, y) \tag{6}$$

However, we do not use $R'_{avg}(x, y)$ (for individual couplets (x, y)) directly for constructing the AcR based distance matrix $D_{R'}$. Its reason is explained with Fig. 2a and b, which show the distributions of R' for two couplets (HOM,TAR) and (MYO,TUR), present in the Mammalian dataset of 37 taxa and 447 gene trees [15, 21]. Figure 2a and b show that $R'_G(x, y)$ values for a couplet (x, y) have highly variable distribution, since individual gene trees $G \in \mathbf{G}$ exhibit high topological incongruence. Many $R'_G(x, y)$ values have very low frequencies. Topologies of the gene trees G containing these $R'_G(x, y)$ values can be considered as infrequent with respect to $\mathbf{G}$. In view of this, we propose a novel filtered averaging (as defined below), by first discarding such infrequent $R'_G(x, y)$ values.

For a couplet (x, y), suppose its set of $R'_G(x, y)$ values ($G \in \tau_{xy}$) are placed in m equal spaced bins. So, width of a particular bin B_i ($1 \le i \le m$) becomes $(\max(R'_G(x, y)) - \min(R'_G(x, y))) / m$. Here, $\max()$ and $\min()$, denote the maximum and minimum operators, respectively. Further, let f^{B_i} be the cardinality of B_i, and $F^B = \max(f^{B_i}) \, \forall \, i$. Our proposed filtered averaging first computes the set R'^m_{xy}, defined below:

$$R'^m_{xy} = \{R'_G(x, y) : R'_G(x, y) \in B_i (1 \le i \le m) \text{ and } f^{B_i} \ge 0.5 \times F^B\}.$$

Algorithm 1. AcRNJXL algorithm

---

1 /* Input: Set of rooted gene trees G.   */
2 /* Output: Rooted species tree S */
3 Initialize S as a star tree with all N taxa covered in **G**.
4 Form $D_{R'}$ and D_X distance matrices, using $R'_G(x,y)$ and $X_G(x,y)$ values for all (x,y) and for all $G \in$ **G**.
5 **while** S *is not fully binary* **do**
6   **for** *All pairs of leaf nodes or* taxa clusters *(defined below)* x *and* y **do**
7   Compute CR'_{xy} using Eq. 8.
8   Compute CX_{xy} using Eq. 7.
9   Choose the couplet (x,y) for coalescence if $(x,y) = argmin_{\forall x',y',x' \neq y'}(CR'_{x'y'} \times CX_{x'y'})$.
10   Create one speciation node n_{xy} (also called a *taxa cluster*) and insert x and y as its children.
11   Set distance (in terms of AcR) from n_{xy} to any other leaf node (or taxa cluster) z as:
12     $d_{R'}(n_{xy},z) = (d_{R'}(x,z) + d_{R'}(y,z))/2$.
13   Set distance (in terms of excess gene leaves) from n_{xy} to z as:
14     $d_X(xy,z) = (d_X(x,z) + d_X(y,z))/2$.
15   Continue agglomeration in successive iterations.

Such 50 % threshold is empirical, and produces best results on biological datasets. Average of the elements in R'^m_{xy} is denoted as $R'^m_{avg}(x,y)$. The distance matrix $D_{R'}$ is constructed by using $R'^m_{avg}(x,y)$ values for all couplets (x,y).

2.3 AcRNJXL Algorithm

We use both D'_R and D_X to construct the species tree S. Initially, S is a star tree covering N taxa covered in **G**. First, for individual couplets (x,y), we compute its *relative XL distance* CX_{xy} and *relative AcR distance* CR'_{xy}, with respect to other couplets (x,z) and (y,z) for all other taxa $z(\neq x,y)$, according to the following equations:

$$CX_{xy} = (N-2)d_X(x,y) - \sum_{z(\neq x,y)}(d_X(x,z) + d_X(y,z)) \tag{7}$$

$$CR'_{xy} = (N-2)d_{R'}(x,y) - \sum_{z(\neq x,y)}(d_{R'}(x,z) + d_{R'}(z,y)) \tag{8}$$

Lower CX_{xy} means both $d_X(x,y)$ is low and $\sum_{z(\neq x,y)}(d_X(x,z) + d_X(y,z))$ is high. Similarly, CR'_{xy} is low when both $d_{R'}(x,y)$ is low, and $\sum_{z(\neq x,y)}(d_{R'}(x,z) + d_{R'}(z,y))$ is high. Both conditions indicate that on average (with respect to **G**), coalescence of x and y has occurred before other couplets. Thus, a couplet (x,y) which is a possible candidate for coalescence at the current iteration, should have low (minimum or close to minimum) CR'_{xy} and CX_{xy} values. So, the product of CR'_{xy} and CX_{xy} should also be very low (minimum or close to minimum). Following this principle, each iteration of our proposed NJ based agglomeration selects a couplet (x, y) for coalescence, provided the following:

$$(x, y) = argmin_{\forall x', y', x' \neq y'} (CR'_{x'y'} \times CX_{x'y'}) \qquad (9)$$

Coalescence of a couplet (x, y) generates new internal speciation node n_{xy}, whose child nodes are x and y. Second step of NJ based agglomeration involves re-adjusting the distance matrices $D_{R'}$ and D_X, by estimating distances between n_{xy} and any other leaf node (or any other internal speciation node) z. Let, $d_{R'}(n_{xy}, z)$ and $d_X(n_{xy}, z)$ denote the approximated AcR and XL measures, respectively, between the pair of nodes n_{xy} and z. These values are computed using simple averaging, as the following:

$$d_{R'}(n_{xy}, z) = (d_{R'}(x, z) + d_{R'}(y, z))/2. \qquad (10)$$

$$d_X(n_{xy}, z) = (d_X(x, z) + d_X(y, z))/2. \qquad (11)$$

Such agglomeration continues until a binary species tree S is generated. The algorithm is referred to as **AcRNJXL** (NJ based species tree estimation with AcR and XL measures). Algorithm 1 presents an overview of AcRNJXL. Performance comparison of AcRNJXL with the reference methods, is provided in Sect. 3.

Computational Complexity: Computation of LCA^G_{xy} for a gene tree G and a couplet $(x, y) \in L(G)$, can be performed at constant time, by applying preorder traversal of all $n \in NL(G)$. Given $M = |\mathbf{G}|$ and N as the total number of taxa covered in $\mathbf{G}$, $R'_G(x, y)$ and $X_G(x, y)$ for all couplets can be computed in $O(MN^2)$ time. NJ [20,23] based agglomeration requires $O(N^3)$ time complexity for N taxa. So, overall time complexity of AcRNJXL is $O(N^3 + MN^2)$, which is equal to that of the couplet based approaches NJst [11], STAR [13], and lower than other reference approaches. As the time complexity is proportional to $O(M)$, AcRNJXL requires lower running time for processing thousands of gene trees. AcRNJXL requires $O(MN^2)$ space complexity, to store couplet based AcR and XL measures for all input trees. Such storage complexity is equal to that of iGTP, Phylonet, STAR, STEAC, and GLASS, and lower than other reference methods such as ASTRAL2, MP-EST, etc.

3   Experimental Results

We have implemented AcRNJXL in python (version 2.7). Phylogenetic library Dendropy (version 3.12.0) [24] is used for reading and processing tree datasets.

Biological Datasets: Following biological datasets are analyzed for performance comparison between AcRNJXL and the reference techniques.

   (1) *Amniota dataset* [3,15] consists of 16 taxa and 248 gene trees. Gene trees were inferred from the nucleotide sequences of respective genes, using maximum

likelihood. Chiari et al. [3] reported a reference species tree by using MP-EST [12] on input gene alignments. We have used gene trees and sequence alignments from the datasets reported in [3,15], and run multi locus bootstrapping (employing both site and gene re-sampling with 200 bootstrap replicates) on individual gene trees using RAxML (version 8.0.17) [22].

(2) Angiosperm dataset [27] concerns about the placement of *Amborella trichopoda Baill* on the evolutionary tree of angiosperms. Study in [26] and output of CA+ML (Concatenated Analysis with Maximum Likelihood) conclude that Amborella is sister to the rest of angiosperms, followed by water lilies (Nymphaeales) [16]. Another study by Xi et al. [27], and output of MP-EST [12], provide an alternative hypothesis that Amborella is sister to water lilies, and together this group is sister to other angiosperms. Considering that MP-EST models ILS (while CA+ML does not model ILS), second hypothesis is more likely for ILS based modeling. We have tested both of these hypotheses using alignments of 310 nuclear genes from 42 angiosperms and 4 outgroups (obtained from http://www.cs.utexas.edu/~phylo/datasets/astral2/). We have also generated 200 bootstrap replicates for individual gene trees.

(3) Mammalian dataset [21] covers 37 taxa and 447 gene trees, with 440 distinct gene tree topologies. Mirarab et al. [15] removed 23 mislabeled genes, and reconstructed gene trees with RAxML [22], using the remaining 424 genes. Input gene trees and alignments were obtained from http://www.cs.utexas.edu/users/phylo/datasets/astral/. Gene trees were rooted using Chicken as the outgroup. Individual gene trees and corresponding alignments were used to estimate 200 bootstrap replicates per gene trees.

Methods Compared: AcRNJXL is benchmarked with the summary methods ASTRAL2 (version 4.7.8) [15,16], STAR [13], NJst [11], and the parsimony approaches Phylonet (version 3.5.6) [25], mulRF (version 1.2) [2], and iGTP (version 1.1) [1]. Methods STEAC [13] and GLASS [17] could not be executed since the coalescence time information was not available. Bayesian methods like BEST [9], *BEAST [6], or BUCKy [8], have not been tested, due to their excessive computation. We have executed ASTRAL2 [16] using its default settings, without using any excess bipartitions generated from MP-EST, STAR, or concatenation analysis. Phylonet [25] was also executed with its default settings, since its exact version incurs huge running time even for the datasets with 20 taxa.

3.1   Performance Comparison on Amniota Dataset

Species tree topologies and bootstrap clade supports for AcRNJXL and other reference methods, when executed on Amniota dataset, are shown in Fig. 3a. This dataset exhibits high degree of conflicts among input gene trees, regarding the position of turtles with respect to the Archosauria clade (consisting of birds and crocodiles). Phylogenetic analysis suggests that the turtles should be placed as a sister group to the Archosauria clade. *Species trees generated from ASTRAL2, NJst, mulRF, STAR, and AcRNJXL, satisfy this configuration.* Both Phylonet and iGTP (their output species trees are shown in the Fig. 3 of the supplementary material) fail to support this clade. *AcRNJXL shows highest (74 %) bootstrap support for this clade, much better than ASTRAL2 (47 %), STAR (49 %), and NJst (38 %), and mulRF (60 %).*

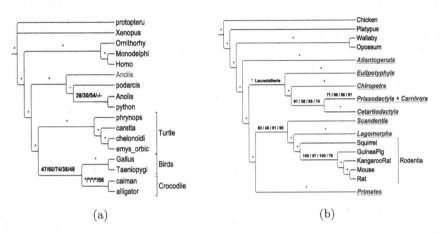

(a)   (b)

Fig. 3. Topology and bootstrap clade supports corresponding to (a) Amniota, and (b) Mammalian dataset, for AcRNJXL and other reference approaches. 100 % Bootstrap support values are indicated by '*'. Bootstrap values for unsupported branches are denoted as '-'. Topology of a group of taxa, depicting 100 % bootstrap, are represented by the group name, shown in colored and underlined labels. (a) For amniota dataset, Both NJst and STAR place Anolis above podarcis and python (shown in red branch; corresponding bootstrap clade s upport is also shown by red colored label.). Rest of the methods place Anolis as a sister to python (shown in green branch). Other branches (shown in black) are supported by all methods. Bootstrap support values (< 100 %) are ordered as ASTRAL2 / mulRF / AcRNJXL / NJst / STAR. (b) For mammalian dataset, Bootstrap support values (< 100 %) are shown in the order of ASTRAL2 / mulRF / STAR / AcRNJXL. NJst performs the same as STAR, thus its topology is not separately shown. (Color figure online)

For gene trees generated from nucleotide sequences, the reference species tree [3] contains the triplet *(podarcis, (Anolis, python))* (in Newick format). On the other hand, reference species tree produced by gene trees derived from amino acid sequences, contains this triplet as *(Anolis, (podarcis, python))*. So, this triplet discriminates between the nucleotide and amino acid based species trees. As we have used gene trees derived from nucleotide sequences, the clade *(podarcis, (Anolis, python))* should be highly supported. However, species trees generated from both NJst and STAR falsely produce the triplet *(Anolis, (podarcis, python))*, with 100 % clade support. Although ASTRAL2 and mulRF correctly predict this triplet, corresponding bootstrap support values are very low (28 % and 30 %, respectively). *AcRNJXL exhibits much higher bootstrap support value (54 %) for this clade. So, AcRNJXL performs best for the Amniota dataset.*

3.2   Performance Comparison on Mammalian Dataset

Figure 3b shows different species tree topologies and bootstrap support values for different species tree construction methods, when executed on the mammalian dataset. Phylonet and iGTP produce wrong species tree topologies within Laurasiatheria clade. Corresponding topologies are shown in the supplementary material (Figs. 5 and 6). NJst produces identical performance with STAR. So its output topology is not shown separately. Outputs of various methods differ mainly with respect to the bootstrap support values, corresponding to the placement of Scandentia, Lagomorpha, and Rodentia groups, as sister to the Primates group. *We see that AcRNJXL exhibits highest bootstrap support (90 %) for this clade, higher than ASTRAL2 (83 %) and much higher than mulRF (40 %) and STAR (61 %).*

3.3   Performance Comparison on Angiosperm Dataset

Species tree obtained by AcRNJXL for Angiosperm dataset is shown in Fig. 4. Trees generated by NJst, STAR, ASTRAL2, and mulRF, are shown in the supplementary material (Figs. 1 and 2). Results depict that ASTRAL2 and mulRF place Amborella as a lone sister to the rest of Angiosperms, followed by Nupher, with bootstrap supports of 75 % and 22 %, respectively. Such result matches with the output of CA+ ML, which does not model ILS. Authors in [16] have argued that the alternative hypothesis (placement of Amborella + Nupher as sister to the rest of the Angiosperms) cannot be recovered by sole ILS based species tree algorithm. However, we have found that NJst, STAR, and AcRNJXL correctly recover such sister relationship, with bootstrap supports of 100 %, 99 %, and 100 %, respectively. Such observation matches with the outputs of MP-EST [12] and the species tree topology reported in [27].

Species tree reported by Xi et al. [27], favors the placement of *monocots* as a sister to *magnollids + eudicots*. However, we have found that only mulRF supports such placement with a relatively high bootstrap support of 58 %. Methods ASTRAL2, NJst and STAR exhibit very low bootstrap supports for this clade. AcRNJXL places *monocots + magnollids* as a sister to *eudicots* with 99 % bootstrap support.

Both NJst and STAR wrongly (according to the studies performed in [26,27]) place the taxon *Vitis* just below *Aquilegia*, with low bootstrap support values of 25 % and 11 %, respectively. Rest of the methods correctly place it above Eucalyptus (as shown in Figs. 1 and 2 of the supplementary material). The method mulRF falsely assigns *Eucalyptus* just above the group *Fagales* (shown by the red colored branch of the Fig. 2 in the supplementary material), with a low bootstrap support of 23 %. Rest of the methods correctly place *Eucalyptus* as a sister clade to *Fagales + Malpighiales*. With respect to such comparative analysis, we can say that AcRNJXL performs best for the Angiosperm dataset.

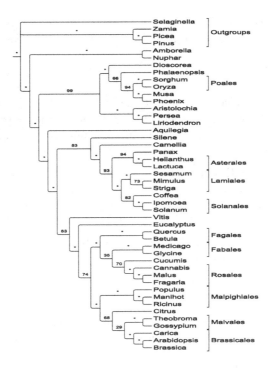

Fig. 4. Topology and bootstrap clade supports for the Angiosperm dataset, for AcRN-JXL. It exhibits 100 % bootstrap support for the placement of (Amborella, Nupher) together, and above the rest of the angiosperms. Here, 100 % bootstrap support values are indicated by '*' symbol.

4   Discussion

We have proposed two new couplet based measures, namely AcR and XL, for species tree construction. AcR measure (R') has higher range of distribution compared to the coalescence rank (R). So, R' discriminates individual couplets better than R. However, higher $R'_G(x,y)$ (for a particular couplet (x,y) in a gene tree G) may not always indicate the closeness of LCA^G_{xy} to the tree root. The measure XL uses tree rooting information, but is also ambiguous like the coalescence rank R. So, sole use of either AcR or XL measures for species tree construction, may produce comparatively lower performances. We have tested both of these approaches and found their comparative lower performances on biological datasets. Using a combination of AcR and XL measures discriminates individual couplets better, and also identifies correct order of evolution among the constituent taxa set. This is evident from the obtained results on biological datasets, presented in Sect. 3.

Parsimony approaches satisfying MDC criterion (tools Phylonet [25], iGTP [1]) exhibit comparatively poor performances. Here, derived species tree S becomes topologically closer to the input gene tree set **G**. However, during ILS, topology of the species tree may be considerably different from the input gene trees. So the derived species tree S may be completely different from the more likely species tree.

Approaches Phylonet and ASTRAL2 incur higher running time for large number (thousands) of input gene trees. AcRNJXL, like other couplet based approaches STAR or NJst, is computationally efficient for such a large number of gene trees. Due to its superior performance on biological datasets, support on incomplete, non-binary gene trees, and computational efficiency, AcRNJXL can be considered as among the best species tree inference methods.

Supplementary and Executable: Supplementary material corresponding to this manuscript, executable and results of AcRNJXL, are provided in http://www.facweb.iitkgp.ernet.in/~jay/phstree/AcRNJXL/AcRNJXL.html.

Acknowledgments. The first author acknowledges Tata Consultancy Services (TCS) for providing the research scholarship.

References

1. Chaudhary, R., Bansal, M.S., Wehe, A., Fernández-Baca, D., Eulenstein, O.: iGTP: a software package for large-scale gene tree parsimony analysis. BMC Bioinform. **23**(574), 1–7 (2010)
2. Chaudhary, R., Burleigh, J.G., Fernández-Baca, D.: Inferring species trees from incongruent multi-copy gene trees using the Robinson-Foulds distance. Algorithms Mol. Biol. **8:28**(1), 1–12 (2013)
3. Chiari, Y., Cahais, V., Galtier, N., Delsuc, F.: Phylogenomic analyses support the position of turtles as the sister group of birds and crocodiles (archosauria). BMC Biol. **10**(65), 1–14 (2012)
4. DeGiorgio, M., Degnan, J.: Robustness to divergence time underestimation when inferring species trees from estimated gene trees. Syst. Biol. **63**(1), 66–82 (2014)
5. Degnan, J.H., Rosenberg, N.A.: Gene tree discordance, phylogenetic inference and the multispecies coalescent. Trends Ecol. Evol. **24**(6), 332–340 (2009)
6. Heled, J., Drummond, A.J.: Bayesian inference of species trees from multilocus data. Mol. Biol. Evol. **27**(3), 570–580 (2010)
7. Kubatko, L.S., Carstens, B.C., Knowles, L.: Stem: species tree estimation using maximum likelihood for gene trees under coalescence. Bioinformatics **25**(7), 971–973 (2009)
8. Larget, B.R., Kotha, S.K., Dewey, C.N., Ané, C.: Bucky: Gene tree/species tree reconciliation with bayesian concordance analysis. Bioinformatics **26**(22), 2910–2911 (2010)
9. Liu, L.: Best: bayesian estimation of species trees under the coalescent model. Bioinformatics **24**(21), 2542–2543 (2008)
10. Liu, L., Xi, Z., Wu, S., Davis, C.C., Edwards, S.V.: Estimating phylogenetic trees from genome-scale data. Ann. N. Y. Acad. Sci. **1360**(1), 36–53 (2015)
11. Liu, L., Yu, L.: Estimating species trees from unrooted gene trees. Syst. Biol. **60**(5), 661–667 (2011)
12. Liu, L., Yu, L., Edwards, S.V.: A maximum pseudo-likelihood approach for estimating species trees under the coalescent model. BMC Evol. Biol. **10**(302), 1–18 (2010)

13. Liu, L., Yu, L., Pearl, D.K., Edwards, S.V.: Estimating species phylogenies using coalescence times among sequences. Syst. Biol. **58**(5), 468–477 (2009)
14. Mirarab, S., Bayzid, M.S., Warnow, T.: Evaluating summary methods for multi-locus species tree estimation in the presence of incomplete lineage sorting. Syst. Biol. p. syu063 (2014). doi:10.1093/sysbio/syu063
15. Mirarab, S., Reaz, R., Bayzid, M.S., Zimmermann, T., Swenson, M.S., Warnow, T.: Astral: genome-scale coalescent-based species tree estimation. Bioinformatics **30**(17), i541–i548 (2014)
16. Mirarab, S., Warnow, T.: ASTRAL-II: coalescent-based species tree estimation with many hundreds of taxa and thousands of genes. Bioinformatics **31**(12), i44–i52 (2015)
17. Mossel, E., Roch, S.: Incomplete lineage sorting: consistent phylogeny estimation from multiple loci. IEEE/ACM Trans. Comput. Biol. Bioinform. **7**(1), 166–171 (2010)
18. Nakhleh, L.: Computational approaches to species phylogeny inference and gene tree reconciliation. Trends Ecol. Evol. **28**(12), 719–728 (2013)
19. Roch, S., Steel, M.: Likelihood-based tree reconstruction on a concatenation of aligned sequence data sets can be statistically inconsistent. Theor. Population Biol. **100**, 56–62 (2015)
20. Saitou, N., Nei, M.: The neighbor-joining method: a new method for reconstructing phylogenetic trees. Mol. Biol. Evol. **4**(4), 406–425 (1987)
21. Song, S., Liu, L., Edwards, S.V., Wu, S.: Resolving conflict in eutherian mammal phylogeny using phylogenomics and the multispecies coalescent model. Proc. Nat. Acad. Sci. USA **109**(37), 14942–14947 (2012)
22. Stamatakis, A.: RAxML-VI-HPC: maximum likelihood-based phylogenetic analyses with thousands of taxa and mixed models. Bioinformatics **22**(21), 2688–2690 (2006)
23. Studier, J.A., Keppler, K.L.: A note on the neighbor-joining algorithm of saitou and nei. Mol. Biol. Evol. **5**(6), 729–731 (1988)
24. Sukumaran, J., Holder, M.T.: DendroPy: a python library for phylogenetic computing. Bioinformatics **26**(12), 1569–1571 (2000)
25. Than, C., Nakhleh, L.: Species tree inference by minimizing deep coalescences. PLOS Comput. Biol. **5**(9), 1–12 (2009)
26. Wickett, N.J., et al.: Phylotranscriptomic analysis of the origin and early diversification of land plants. Proc. Nat. Acad. Sci. USA **111**(45), E4859–E4868 (2014)
27. Xi, Z., Liu, L., Rest, J.S., Davis, C.C.: Coalescent versus concatenation methods and the placement of amborella as sister to water lilies. Syst. Biol. **63**(6), 919–932 (2014)
28. Yu, Y., Warnow, T., Nakhleh, L.: Algorithms for MDC-based multi-locus phylogeny inference: beyond rooted binary gene trees on single alleles. J. Comput. Biol. **18**(11), 1543–1559 (2011)

Bootstrapping Algorithms for Gene Duplication and Speciation Events

Agnieszka Mykowiecka$^{(\boxtimes)}$ and Pawel Górecki

Faculty of Mathematics, Informatics and Mechanics,
University of Warsaw, Warsaw, Poland
{a.mykowiecka,gorecki}@mimuw.edu.pl

Abstract. Based on the classical non-parametric bootstrapping for phylogenetic trees, we propose a novel bootstrap method to define support for gene duplication and speciation events. While this approach can be used to annotate orthology and paralogy, we show how it can be used to verify reliability of tree reconciliation with applications to the problem of rooting of an unrooted gene tree. We propose a linear time algorithm for the computation of bootstrap values and we show the correspondence of our method with the classical non-parametric bootstrapping. Finally, based on simulated data and nine yeast genomes we present a comparative study of tree rooting methods and evaluation of their performance by using our bootstrapping method. The software and examples are publicly available.

Keywords: Non-parametric bootstrapping · Tree reconciliation · Gene duplication · Speciation · Gene tree · Species tree

1 Introduction

The study on the evolutionary history of genes and species is of great importance for many disciplines such as ecology, biology or medicine. Earlier research was severely limited due to difficulties in acquiring DNA sequences of genes and the whole genomes. Today, thanks to the development of new methods of DNA sequencing and conducted large-scale metagenomic experiments many data are publicly available for the purpose of phylogenetic studies.

Examining relations between aligned molecular sequences from a single gene family allows for approximation of the evolutionary history that can be represented in the form of a phylogenetic tree, called *gene tree*. In many cases tree-building methods give slightly different results, and none of them guarantees the correct topology of the inferred gene tree. Thus, the necessary step is to assess the credibility of inferred trees. It can be done by bootstrap, which allows to determine whether a particular tree is a good approximation of the evolution of molecular sequences [9]. *Species trees* that represent evolutionary history of species can be inferred from gene trees. However, gene and species trees are usually incongruent which can be due to data selection, sequencing errors, inference

© Springer International Publishing Switzerland 2016
M. Botón-Fernández et al. (Eds.): AlCoB 2016, LNBI 9702, pp. 106–118, 2016.
DOI: 10.1007/978-3-319-38827-4_9

methods or evolutionary events such as gene duplication, loss or gene transfer [20]. Studies on gene and species phylogeny have been conducted since 1980s. Goodman et al. [11] introduced a model of *tree reconciliation* in which gene duplication and loss events are invoked to address the differences between a gene tree and its species tree. This concept was later formalized [27] by introducing reconciled trees and the *duplication-loss cost*, i.e., the minimal number of gene duplications and losses required to reconcile a given gene tree with its species tree. The model was further extended by horizontal gene transfer [20, 24].

A major problem in tree reconciliation is its high sensitivity to fallible gene trees [19, 31]. Gene trees can have topological errors that result in an incorrect topology or rooting errors related to the wrong placement of the root [12]. One way to deal with errors is the reconciliation model in which input gene trees are corrected by using tree edit operations [5, 6]. For example, in [5, 12] a gene tree have one error if there is a tree with improved reconciliation cost in the local neighborhood of the given gene tree. Another method proposed in [3, 32] consisted in preprocessing of a set of gene trees by removing nodes that cause inconsistency. A more statistical approach was presented in [34].

While the classical reconciliation model is applicable to rooted trees only, most of the standard tree inference methods infer unrooted trees. In addition it is often difficult to identify a credible rooting. Outgroup rooting can result in incorrect rootings when there is heterogeneity in a gene tree. Moreover, when analyzing an ancient dataset or a gene family that exists only in a specific organismal group, the outgroup can be unavailable [10, 23]. Another approach is to root gene trees under the molecular clock assumption, or similarly by using midpoint method. Both can result in error when there is a molecular rate variation throughout the tree [21, 22]. Tree reconciliation has been successfully extended to reconcile an unrooted gene tree with a rooted species tree by seeking a rooting of the unrooted gene tree that invokes the minimal duplication-loss cost [16, 35].

So far, bootstrap methods in tree reconciliation were mainly focused on rooted trees. For example, [28] proposed to estimate the support of horizontal gene transfers. An early approach to integrate reconciliation and bootstrapping for unrooted trees [36] relied on rooting by choosing the midpoint rooting with the minimal duplication-loss cost. By aggregating gene duplication locations from all rooted sample trees, the authors of [36] were able to compute support values for every duplication from the input gene tree. However, in such an approach, some information can be lost as the selected midpoint rooting may be incorrect.

Here we propose to assess the credibility of unrooted gene trees by using unrooted reconciliation and non-parametric bootstrapping. Having this, we show how to compute support values for branching events in a gene tree, i.e., for duplication and speciation events. We propose a linear time algorithm for the computation of bootstrap values and we show the correspondence of our approach with the classical non-parametric bootstrapping. Finally, based on simulated data and yeast genomes we present a comparative study of tree rooting methods.

2   Basic Definitions

We introduce several notions from the phylogenetic theory [15,27]. For a tree T by L_T we denote the set of all leaves present in T. A *species tree* is a rooted binary tree whose leaves are called *species*. A *rooted gene tree* T over a species tree S is a triple $\langle V_T, E_T, \Lambda_T \rangle$ such that $\langle V_T, E_T \rangle$ is a rooted binary tree and $\Lambda_T \colon L_T \rightarrow L_S$ is the leaf labelling function called *labelling*. Leaves of a gene tree are called *genes*. A *cluster* for v is the set of all leaves present in the subtree of T rooted at v. We denote trees by using the standard nested parenthesis notation with the extension that allows to encode labelling. For instance, in Fig. 1A, $G = ((a1, d), (a2, b))$ is a four-leaf gene tree over a species tree $(((a, b), c), d)$ such that two leaves of the gene tree, i.e., $a1$ and $a2$, are assigned to species a.

Let T be a rooted gene tree over a species tree S. Let $\mathsf{lca}_S(v, w)$ denote the least common ancestor of nodes v and w from S. *The least common ancestor mapping*, or *lca-mapping*, $\mathsf{M} \colon V_T \rightarrow V_S$, is defined as follows: $\mathsf{M}|_{L_T} = \Lambda_T$ and when $v \in V_T$ has two children a and b, then $\mathsf{M}(v) = \mathsf{lca}_S(\mathsf{M}(a), \mathsf{M}(b))$. We distinguish two *types* of nodes: (I) an internal node v is a *duplication* if $\mathsf{M}(g) = \mathsf{M}(a)$ for a child a of g, and (II) *speciation* nodes, otherwise. *The duplication cost*, denoted by $\mathsf{D}(T, S)$, is the total number of duplications in T [26]. The total number of *gene losses* required to reconcile T and S can be defined by: $\mathsf{L}(T, S) = 2\,\mathsf{D}(T, S) + \sum_{g \text{ is internal}, a, b \text{ children of } g}(\|\mathsf{M}(a), \mathsf{M}(b)\| - 2)$, where $\|a, b\|$ is the number of edges on the path connecting a and b in S [24]. Finally, we can define the *duplication-loss cost*: $\mathsf{DL}(T, S) = \mathsf{D}(T, S) + \mathsf{L}(T, S)$.

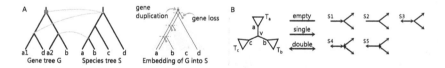

Fig. 1. *A: An example of rooted reconciliation.* The lca-mapping between a gene tree G and a species tree S and the embedding of G into S, i.e., an informal representation of an evolutionary scenario explaining differences between G and S by using gene duplications and gene losses. Here the DL cost is 5 (1 duplication + 4 gene losses). The lca-mapping is shown only for the internal nodes. *B: Stars in unrooted reconciliation.* A star in a gene tree and possible types of edges and stars. Subtrees T_a, T_b and T_c are rooted at a, b and c, respectively.

Now we introduce the main concepts related to unrooted reconciliation. For a species tree S, *an unrooted gene tree* G over a S is a triple $\langle V_G, E_G, \Lambda_G \rangle$ such that $\langle V_G, E_G \rangle$ is an undirected acyclic connected graph in which each node has degree 1 (leaves) or 3 (internal nodes), and $\Lambda_G \colon L_G \rightarrow L_S$ is a leaf labelling function. A *split* $A|B$ is a partition of X, i.e., A and B are two disjoint non-empty sets such that $A \cup B = X$. We say that a split $A|B$ is present in an unrooted tree if there is an edge e in G, such that removing e from G induces two subtrees of G having A and B as the set of its leaves, respectively. Splits are the unrooted equivalent of clusters.

For an edge $e \in E_G$, by G_e, we denote the *rooting*, i.e., a rooted gene tree, obtained from G by placing the root on e. By M_e we denote the lca-mapping between G_e and S. For a species tree S, such a rooting induces the duplication-loss cost $DL(G_e, S)$. The set of all edges with the minimal duplication-loss cost, or *optimal edges*, we call *plateau*. Rootings of optimal edges are also called *optimal*.

Without loss of generality we assume that every root of a gene tree is mapped into the root of S, denoted by $\top$, and both trees are non-trivial. An edge $e = \{v, w\}$ of G is *empty* if the root of G_e is a speciation. We call e *double* if $M_e(v) = \top = M_e(w)$. Otherwise, e is called *single*. A single edge e is called *v-incoming* or *w-outgoing* if $M_e(v) \neq \top = M_e(w)$. Let v be an internal node of G, then a *star* with a *center* v consists of three edges sharing a common node v incident to nodes a, b and c, respectively (see Fig. 1B). There are five types of possible star topologies: the $S1$ star has one v-incoming edge and two v-outgoing edges, the $S2$ star has exactly two v-outgoing edges and one empty edge, the $S3$ star has two v-outgoing edges and one double edge, in the $S4$ star all 3 edges are double, and the $S5$ star has one v-outgoing edge and two double edges. The main result on unrooted reconciliation is below.

Theorem 1 (Adopted from [16]). *For a given unrooted gene tree G, we have: (1) either G has exactly one empty edge or G has at least one double edge, (2) if the plateau of G consists of exactly one edge then this edge is either empty or double, and all other edges are single, or (3) if the plateau of G has more than one edge then it contains all edges present in stars $S4$ and $S5$, and all other edges are single.*

It follows from this theorem and the properties of stars that the plateau is a full subtree of G. See also [15,16] for more details. An example is depicted in Fig. 2.

3   Results

In this Section we present theoretical results related to bootstrapping with reconciliation. We start with the properties of optimal rootings, then we define the main notion of support values for duplication and speciation events. Next, we propose an algorithm for the support values computation. Finally, we show a correspondence between our approach and the classical bootstrapping proposed by Felsenstein [9].

3.1   Properties of Optimal Rootings

Lemma 2. *If e and e' are elements of the plateau of an unrooted gene tree G, then $M_e(w) = M_{e'}(w)$ for any $w \in V_G$.*

Note that the root of any rooting of unrooted gene tree is mapped to the same node in S. Despite the formal complication with the root (which is formally a new node in a rooting of unrooted tree), we conclude that lca-mappings of optimal rootings are identical. The next result follows from Lemma 2.

Theorem 3 (Homology in unrooted trees). *For a node of an unrooted gene tree G, its type is the same in every optimal rooting of G.*

Now we introduce a notion of *type* of cluster in unrooted trees. A cluster of a node v from a rooting of G is called a *duplication* if v is a duplication in an optimal rooting of G (or, by Theorem 3, equivalently in all optimal rootings). Analogically, we define a *speciation cluster* in G.

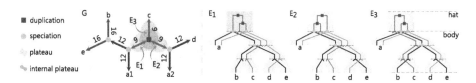

Fig. 2. *Left*: An example of unrooted gene tree G, reconciled with $S = (a, ((b, c), (d, e)))$, with three optimal rootings. Each edge is decorated with the DL cost of the corresponding rooting (optimal cost is 9). G has three non-leaf speciation clusters (marked by green circles) and two duplication clusters (one marked by blue square + the cluster of the root). Every plateau edge, colored in red, has a label E1, E2 or E3, which relates to one of the embeddings visible on the left. *Right*: Three embeddings of all optimal rootings of G into S. Every embedding has 2 duplications and 7 gene losses.(Color figure online)

Lemma 4. *There are four disjoint kinds of speciation and duplication clusters in unrooted gene trees: (1) clusters of internal nodes of the plateau, (2) clusters of leaves of the plateau, (3) clusters of nodes disjoint with the plateau, and (4) the rooting cluster composed of all leaves. The last three kinds are present in every optimal rooting.*

Optimal evolutionary scenarios can be represented by embeddings of an optimal rooting into the species tree [14]. From Lemma 2 and Theorem 3, we conclude that these scenarios differ only in their rooting edges (*hat*) while the remaining parts of all the trees (*body*) are identical as indicated in Fig. 2 (see also [16]).

3.2   Bootstrapping

Bootstrap methods are used to assign confidence values for the estimated trees. Bootstrapping consists in reconstruction of trees altered by random input data modification or sampling. The variability in the obtained set is assessed by comparing its elements with the original tree. The results can be interpreted as an indication of the influence of arbitrary changes which (in this case) do not resemble evolutionary schema on the structure of the phylogenetic tree.

Our methods is partially based on the classical non-parametric bootstrap proposed by Felsenstein [9]. Given a set X of n gene sequences and a multiple sequence alignment A (of dimension n rows and k columns) of sequences from X. First, N *bootstrap alignments* are constructed, where each bootstrap alignment is formed by randomly selecting k columns from A with replacement. Next, for each bootstrap alignment, an unrooted gene tree, called *sample tree*, is inferred by using some standard tree-building tool, e.g. PhyML [17]. Finally, for further processing a gene tree G is inferred from the alignment A. The frequency of clusters/splits present in sample trees indicates the support for the corresponding clusters/splits in G.

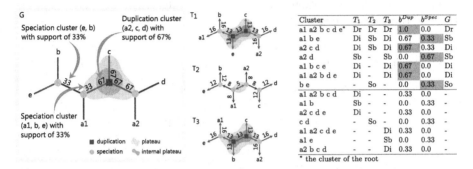

Cluster	T_1	T_2	T_3	b^{Dup}	b^{Spec}	G
a1 a2 b c d e*	Dr	Dr	Dr	1.0	0.0	Dr
a1 b e	Di	Sb	Di	0.67	0.33	Sb
a2 c d	Di	Sb	Di	0.67	0.33	Di
a2 d	Sb	-	Sb	0.0	0.67	Sb
a1 b c e	Di	-	Di	0.67	0.0	Di
a1 a2 b d e	Di	-	Di	0.67	0.0	Di
b e	-	So	-	0.0	0.33	So
a1 a2 b c d	Di	-	-	0.33	0.0	-
a1 b	Sb	-	-	0.0	0.33	-
a2 c d e	Di	-	-	0.33	0.0	-
c d	-	So	-	0.0	0.33	-
a1 a2 c d e	-	-	Di	0.33	0.0	-
a1 e	-	-	Sb	0.0	0.33	-
a2 b c d	-	-	Di	0.33	0.0	-

* the cluster of the root

Fig. 3. *A bootstrap example (see Fig. 2). Left*: a gene tree G with D/S support values shown for non-leaf clusters present in optimal rootings. *Middle*: trees T_1, T_2 and T_3 - sampled from G. Edges of T_i's are decorated with the rooting cost (DL). *Right*: D/S support values for all non-leaf clusters from rootings of G, T_1, T_2 and T_3. Cluster type is denoted by S (speciation) and D (duplication). Additionally, r (root), i (plateau internal), b (plateau border) and o (outside plateau) denote the location of the cluster. For example, the duplication cluster $\{a2, c, d\}$ from G is present as a duplication inside of the plateau of T_1, which is denoted Di, while the same cluster determines a speciation located on the border of the plateau in T_2 (denoted by Sb).

Based on the non-parametric bootstrapping we provide the main notion of duplication and speciation (D/S) support values.

Definition 5 (D/S support values). *Given: a species tree S and a collection of sample trees $\mathcal{U}$ such that all trees from $\mathcal{U}$ have the same set of leaves X (a set of genes) and the same labelling[1] $\Lambda: X \rightarrow L_S$. Then, for a cluster $A \subseteq X$, the duplication support for A is defined as $b^{Dup}(A, \mathcal{U}) = \frac{1}{|\mathcal{U}|}|\{T \in \mathcal{U} : A \text{ is a duplication cluster in } T\}|$, and the speciation support as $b^{Spec}(A, \mathcal{U}) = \frac{1}{|\mathcal{U}|}|\{T \in \mathcal{U} : A \text{ is a speciation cluster in } T\}|$.*

Similarly to the standard non-parametric bootstrapping of phylogenetic trees it is interesting to analyse the support values for the clusters of the gene tree inferred from the input alignment. An example is depicted in Fig. 3.

Problem 6. Given: a species tree S, a gene tree G over S and a collection of sample trees $\mathcal{U}$ such that all trees from $\mathcal{U} \cup \{G\}$ have the same set of leaves X

---

[1] Note that in this definition all trees in $\mathcal{U}$ are over S.

and the same labelling $\Lambda\colon X \to L_S$. For each duplication and speciation cluster A in G compute: $\sigma_G(A,\mathcal{U}) = \begin{cases} b^{Dup}(A,\mathcal{U}) & \text{if } A \text{ is a duplication cluster in } G, \\ b^{Spec}(A,\mathcal{U}) & \text{if } A \text{ is a speciation cluster in } G. \end{cases}$

For a set of edges E_G in G, let $\hat{E}_G = \{\langle v, w\rangle, \langle w, v\rangle\colon \{v, w\} \in E_G\}$. For a directed edge $\langle v, w\rangle \in \hat{E}_G$, by $c(v, w, G)$ we denote the cluster of v in the rooting $G_{\langle v,w\rangle}$. There is one-to-one correspondence between clusters and directed edges, therefore, due to computational efficiency in our algorithm we assign support values to the directed edges only.

Lemma 7. *Under the notation from Algorithm 1. For each duplication and speciation cluster A in G such that $c(v, w, G) = A$ we have $\sigma_G(A,\mathcal{U}) = \#(v, w)/|\mathcal{U}|$.*

Proof (Sketch). We have the following meaning of functions in line 10 of Algorithm 1: $m(v, w, P, Q)$ - the lca-mapping $M_{v,w}(v)$; $\tau(v, w, P, Q)$ - the type of a cluster $c(v, w, Q)$, i.e., duplication, speciation or not present in any optimal rooting; `incoming`(v, w) - the edge $\{v, w\}$ is v-incoming in the star with center v; `insideplateau`(v) - v is internal in the plateau; `symmetric`(v, w) - the edge $\{v, w\}$ is symmetric; and `inoptrooting`(v, w) - the cluster $c(v, w, Q)$ is present in an optimal rooting. The main part is in lines 5–8 in which we increase the counter of events $\#(v, w)$ when the cluster $c(v, w, T_i)$ is present in G. We use efficient lca-queries between R and T_i'''s with additional verification of types of clusters.

By using linear time preprocessing [2], we can prove the next result.

Theorem 8. *Algorithm 1 computes D/S support values in linear time.*

3.3   Correspondence to Classical Bootstrap

Now we present the correspondence between D/S support values and the support values from Felsenstein's bootstrapping of unrooted and rooted gene trees proposed by [9]. We use the notation from Sect. 3.2 and Definition 5. For a split $A|B$ the support for $A|B$ in $\mathcal{U}$ is defined by $s^u(A|B,\mathcal{U}) = \frac{1}{|\mathcal{U}|}|\{T \in \mathcal{U}\colon A|B \text{ is a split in } T\}|$.

Theorem 9. *For a collection of unrooted gene trees $\mathcal{U}$ and a split $A|B$, we have $s^u(A|B,\mathcal{U}) \geq b^{Dup}(A,\mathcal{U}) + b^{Spec}(A,\mathcal{U})$.[2]*

Similarly, we define the support for a cluster A in a collection of rooted trees $\mathcal{R}\colon s^r(A,\mathcal{R}) = \frac{1}{|\mathcal{R}|}|\{T \in \mathcal{R}\colon A \text{ is a cluster in } T\}|$. The D/S support values can be naturally extended to collections of rooted gene trees by replacing the term "cluste" with "node" in Definition 5. We omit the straightforward definitions.

Theorem 10. *For a collection of rooted gene trees $\mathcal{R}$ over the same set of leaves, and a cluster A, we have $s^r(A, \mathcal{R}) = b^{Dup}(A, \mathcal{R}) + b^{Spec}(A, \mathcal{R})$.*

---

[2] Observe that B is not present in the right side of the inequality.

Algorithm 1. Computing D/S Support Values

1: **Input/output:** See Problem 1. Let $\mathcal{U} = \{T_1, T_2, \ldots, T_N\}$.
2: Fix $\omega \in X$. Let $R := G_e$, where e is the edge incident to ω. For $g \in V_G$, let $\pi(g)$ denote the parent of g in R if it is not the root of R, otherwise $\pi(g)$ is the sibling of g. Note that $\pi(g)$ is an element of V_G and $\{g, \pi(g)\} \in E_G$. For each i, let T_i' be the unrooted gene tree over R obtained from T_i by replacing the labelling with the identity function on X.
3: Init lca-structures for S and R. For $\langle v, w \rangle \in \hat{E}_G$, $\#(v, w) := 0$ // reset cluster counters
4: **For** each $i \in 1, 2, \ldots, N$ and **For** each $\langle v, w \rangle \in \hat{E}_{T_i}$ such that $\tau(v, w, T_i, S) \neq None$
5:   **If** $\omega \in c(v, w, T_i)$ **Then** { $g := m(v, w, T_i', R)$
6:     **If** $|c(v, w, T_i)| = |c(g, \pi(g), G)|$ AND $\tau(v, w, S, T_i) = \tau(g, \pi(g), G, S)$ **Then** $\#(v, w) + +$ }
7:   **Else** { $g := m(w, v, R, T_i')$
8:     **If** $|c(w, v, T_i)| = |c(g, \pi(g), G)|$ AND $\tau(v, w, T_i, S) = \tau(\pi(g), g, G, S)$ **Then** $\#(v, w) + +$ }
9: **Return** $\#(v, w)/|\mathcal{U}|$ for each $\langle v, w \rangle \in \hat{E}_G$ such that $\tau(v, w, G, S) \in \{Dup, Spec\}$
10: *Definitions.* For a rooted tree Q, an unrooted gene tree P over Q and $\langle v, w \rangle$ in $\hat{E}_P$:

$$m(v, w, P, Q) := \begin{cases} \Lambda_P(v) & v \text{ is a leaf in } P, \\ \mathrm{lca}_Q(m(x, v, P, Q), m(y, v, P, Q)) & v \text{ is internal and } \{x, y\} = ch(v, w). \end{cases}$$

$$\tau(v, w, P, Q) := \begin{cases} None & \text{not inoptrooting}(v, w, P, Q), \\ Dup & \text{inoptrooting}(v, w, P, Q), \{x, y\} = ch(v, w) \text{ and} \\ & m(v, w, P, Q) = m(x, v, P, Q) \text{ or } m(v, w, P, Q) = m(y, v, P, Q), \\ Spec & \text{otherwise} \end{cases}$$

where for an internal node $v \in V_P$, $ch(v, w) = \{x, y\}$ such that $\{x, y, w\}$ is the set of all neighbours of v; $\top$ is the lowest node in Q whose cluster contains $\Lambda_P(L_Q)$; incoming$(v, w) := m(v, w, P, Q) \neq \top = m(w, v, P, Q) = \top$; symmetric$(v, w) := m(v, w, P, Q) = \top = m(w, v, P, Q) = \top$ OR $m(v, w, P, Q) \neq \top \neq m(w, v, P, Q)$; insideplateau$(v) := \exists$ siblings x and y of v such that: $x \neq y$ AND symmetric(v, x) AND symmetric(v, y) and inoptrooting$(v, w) :=$ incoming(v, w) OR insideplateau(v) OR symmetric(v, w).

---

4   Experiment

We performed several computational experiments with bootstrapping and reconciliation on simulated and real data.

Simulated data preparation. In the first step, *model species trees* were generated using Mesquite [25], with topology generation performed according to the Yule-Harding distribution. The procedure is similar to the one proposed in [4] with tree height set to 115 myr, and the number of leaves equal 16.

Simulated gene trees were created from model species trees using a continuous time birth-death process [1] with the gene duplication and gene loss events. On each lineage, an occurrence of gene duplication (bifurcation) or

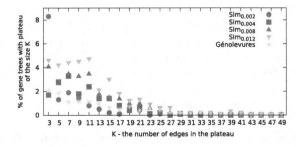

Fig. 4. Frequency diagram of plateau sizes. The numbers of gene trees having singleton plateau (for $K = 1$), omitted here, are present in the second row of Table 1.

loss (termination) was drawn with a probability defined by a constant rate. As duplication should not change the height of a tree, a duplication node was added precisely at the point of the model tree edge in which a duplication event was postulated. In [30], three different values of rates of duplication and losses were proposed: 0.002, 0.004 and 0.008 events/gene per myr. For greater diversity of gene trees, in our experiment we additionally tested the rate of 0.012. For each simulated model tree, 1000 simulated gene trees were generated. For each of them we simulated a nucleotide sequence alignment of length 100 under the GTR + Gamma + I model using Seq-Gen [29]. Next, for each parameter rate $\lambda \in \{0.002, 0.004, 0.008, 0.012\}$, we obtained a set Sim_λ consisting of 1000 *unrooted gene family trees* inferred by *PhyML* program [18] from the corresponding alignments. Finally, from each Sim_λ, we inferred a species tree S_λ by using the program fasturec [13].

Real data preparation. We downloaded the set of 9 yeast genomes consisting of 4617 protein families from [33]. After removing families with only two genes, we inferred 4141 gene trees by using *PhyML* with the standard parameter setting.

Plateau sizes for all datasets are depicted in Fig. 4.

Bootstrap processing. Next steps were performed for all datasets. For each alignment we created 100 bootstrap alignments by *Seqboot* from *PHYLIP* package [8]. Finally, for each bootstrap alignment we inferred a sample tree by *PhyML*.

4.1 A Comparative Study of Rooting Methods

In our study we compared five rooting methods by using the rooting score based on the D/S support values as follows. Given an optimal edge e from a gene tree G and a set of sample trees $\mathcal{U}$, a *rooting score for e* is the average value of $\sigma_G(A, \mathcal{U})$ for all non trivial (non leaf/root) clusters A from G_e. We claim that the edges from the plateau having the maximal rooting score are the best candidates for rooting. We need two additional definitions. The *edge distance* between two nodes is the number of edges on the shortest path connecting these nodes. In the case when the gene tree has branch lengths, the *BL-distance* between two nodes is the total branch length of all edges on the shortest path connecting these nodes. We have three types of standard rooting methods. Two of them take into consideration all tree edges [7] while the last one uses only edges included in the plateau [36]. In **midpoint edge rooting** the root is placed in a half-way between two the most edge distant leaves, while in **BL-Midpoint rooting** the root is placed such that its BL-distance to the leaves is minimized. In **Midpoint plateau rooting** the root is placed in a half-way between two the most distant nodes from the plateau.

Note that in our model of binary trees, the midpoint rootings may be non unique. For instance, if an unrooted tree has three leaves a, b and c, then the midpoint edge rooting can be $(a, (b, c))$, $(b, (a, c))$ or $(c, (b, a))$. The same property holds for the BL-midpoint rootings. Additionally for a control we tested two random rootings: **random edge rooting** and **random plateau rooting** where the root is placed on the edge uniformly chosen from the set of all edges of a gene tree and all edges of the plateau, respectively.

The summary of results is depicted in Table 1, where A denotes the number of gene trees from a given dataset having rooting inside DL-plateau. B is the number of rootings having the maximal rooting score, and C is the percent of gene trees with non-singleton plateau having the maximal rooting score. In case of ambiguity, we assume a match if there is a non-empty intersection between sets of corresponding rootings. Our results suggests that the midpoint edge and BL-midpoint rooting methods indicate generally poorly supported rootings for the simulated datasets. Even the random edge rooting method performs better than these two methods. This observation partially holds for the yeast dataset with the difference that BL-midpoint rootings are generally better supported (1282 well supported rootings).

Table 1. Summary of rootings of gene trees from simulated and real datasets.

	$Sim_{0.002}$			$Sim_{0.004}$			$Sim_{0.008}$			$Sim_{0.012}$			Génolevures		
Dataset size	1000			1000			1000			1000			4141		
# singleton plateau trees	843			830			751			645			3601		
Rooting method	A	B	C	A	B	C	A	B	C	A	B	C	A	B	C
Midpoint edge	101	2	1%	147	14	2%	202	17	3%	302	19	3%	726	88	6%
BL-midpoint	62	13	6%	135	35	8%	180	45	6%	270	52	7%	1407	1282	65%
Midpoint plateau	1000	934	58%	1000	856	15%	1000	800	20%	1000	707	17%	4141	3847	46%
Random edge	131	74	10%	162	81	8%	193	75	10%	263	69	6%	1121	807	9%
Random plateau	1000	884	24%	1000	845	9%	1000	779	11%	1000	693	14%	4141	3660	11%

For the plateau based methods, the number of well supported rootings is usually high due to the large number of singleton plateaux present in our datasets. For example, in the dataset $X_{0.002}$, 843 out of 1000 trees have a unique rooting candidate in the plateau. Therefore, to compare these methods we analyzed non-singleton plateaux (see columns C). In the first simulated dataset the ratio of optimal bootstrap rootings is 58 % for the midpoint plateau rootings. This property can be explained by the fact that relatively large portion of trees has the plateau of size 3 (see Fig. 4). In consequence, in such a case the midpoint plateau rooting method gives all three possible rootings which includes the rooting maximal score. Next, the first dataset performed better than the other simulated datasets, which is due to usually more complex plateaux as indicated in Fig. 4. In the real dataset the midpoint plateau method inferred 46 % rootings with maximal score. However, a better ratio for the non-singleton plateau trees was obtained for the BL-midpoint method. On the other hand the latter method performed poorly for the trees with singleton plateaux.

5   Conclusions and Future Work

In this article we proposed a bootstrapping approach to define support for gene duplication and speciation events when reconciling a given gene tree with its species tree. By comparing gene trees obtained by bootstrapping to the original gene tree we showed how to calculate support for both clusters and gene duplication events. While this approach can be used to annotate orthology and paralogy in unrooted trees, we showed how it can be used to verify reliability of tree reconciliation with applications to the rooting problem. We provided several theoretical and algorithmic results, in particular we showed the correspondence between our method and the classical non-parametric bootstrapping. Software and example are publicly available at: http://bioputer.mimuw.edu.pl/~agnieszka/bootstrap.

Acknowledgements. This work was supported by the DSM funding for young researchers of the Faculty of Mathematics, Informatics and Mechanics of the University of Warsaw.

References

1. Arvestad, L., Berglund, A., Lagergren, J., Sennblad, B.: Gene tree reconstruction and orthology analysis based on an integrated model for duplications and sequence evolution. In: RECOMB, pp. 326–335 (2004)
2. Bender, M.A., Farach-Colton, M.: The LCA problem revisited. In: Gonnet, G.H., Viola, A. (eds.) LATIN 2000. LNCS, vol. 1776, pp. 88–94. Springer, Heidelberg (2000)
3. Beretta, S., Dondi, R.: Gene tree correction by leaf removal and modification: tractability and approximability. In: Beckmann, A., Csuhaj-Varjú, E., Meer, K. (eds.) CiE 2014. LNCS, vol. 8493, pp. 42–52. Springer, Heidelberg (2014)
4. Chaudhary, R., Boussau, B., Burleigh, J.G., Fernandez-Baca, D.: Assessing approaches for inferring species trees from multi-copy genes. Syst. Biol. **64**(2), syu128 (2014)
5. Chaudhary, R., Burleigh, J.G., Eulenstein, O.: Efficient error correction algorithms for gene tree reconciliation based on duplication, duplication and loss, and deep coalescence. BMC Bioinform. **13**(Suppl. 10), S11 (2012)
6. Durand, D., Halldórsson, B.V., Vernot, B.: A hybrid micro-macroevolutionary approach to gene tree reconstruction. J. Comput. Biol. **13**(2), 320–335 (2006)
7. Farris, J.S.: Estimating phylogenetic trees from distance matrices. Am. Nat. **106**(951), 645–668 (1972)
8. Felsenstein, J.: PHYLIP. http://evolution.genetics.washington.edu/phylip.html
9. Felsenstein, J.: Confidence limits on phylogenies: an approach using the bootstrap. Evol. **39**, 783–791 (1985)
10. Gogarten, J.P., Kibak, H., Dittrich, P., Taiz, L., Bowman, E.J., Bowman, B.J., Manolson, M.F., Poole, R.J., Date, T., Oshima, T.: Evolution of the vacuolar h+-atpase: implications for the origin of eukaryotes. Proc. Nat. Acad. Sci. **86**(17), 6661–6665 (1989)
11. Goodman, M., Czelusniak, J., Moore, G.W., Romero-Herrera, A.E., Matsuda, G.: Fitting the gene lineage into its species lineage, a parsimony strategy illustrated by cladograms constructed from globin sequences. Syst. Zool. **28**(2), 132–163 (1979)
12. Górecki, P., Eulenstein, O.: Algorithms: simultaneous error-correction and rooting for gene tree reconciliation and the gene duplication problem. BMC Bioinform. **13**(Suppl. 10), S14 (2012)

13. Górecki, P., Burleigh, J.G., Eulenstein, O.: GTP supertrees from unrooted gene trees: linear time algorithms for nni based local searches. In: Bleris, L., Măndoiu, I., Schwartz, R., Wang, J. (eds.) ISBRA 2012. LNCS, vol. 7292, pp. 102–114. Springer, Heidelberg (2012)

14. Górecki, P., Tiuryn, J.: DLS-trees: a model of evolutionary scenarios. Theoret. Comput. Sci. **359**(1–3), 378–399 (2006)

15. Górecki, P., Eulenstein, O., Tiuryn, J.: Unrooted tree reconciliation: a unified approach. IEEE/ACM Trans. Comput. Biol. Bioinf. **10**(2), 522–536 (2013)

16. Górecki, P., Tiuryn, J.: Inferring phylogeny from whole genomes. Bioinformatics **23**(2), e116–e122 (2007)

17. Guindon, S., Delsuc, F., Dufayard, J., Gascuel, O.: Estimating maximum likelihood phylogenies with PhyML. Methods Mol. Biol. **537**, 113–137 (2009)

18. Guindon, S., Gascuel, O.: A simple, fast, and accurate algorithm to estimate large phylogenies by maximum likelihood. Syst. Biol. **52**(5), 696–704 (2003)

19. Hahn, M.W.: Bias in phylogenetic tree reconciliation methods: implications for vertebrate genome evolution. Genome Biol. **8**(7), R141+ (2007)

20. Hallett, M.T., Lagergren, J.: Efficient algorithms for lateral gene transfer problems. In: RECOMB, pp. 149–156 (2001)

21. Holland, B., Penny, D., Hendy, M.: Outgroup misplacement and phylogenetic inaccuracy under a molecular clock - a simulation study. Syst. Biol. **52**, 229–238 (2003)

22. Huelsenbeck, J.P., Bollback, J.P., Levine, A.M.: Inferring the root of a phylogenetic tree. Syst. Biol. **51**(1), 32–43 (2002)

23. Iwabe, N., Kuma, K., Hasegawa, M., Osawa, S., Miyata, T.: Evolutionary relationship of archaebacteria, eubacteria, and eukaryotes inferred from phylogenetic trees of duplicated genes. Proc. Nat. Acad. Sci. **86**(23), 9355–9359 (1989)

24. Ma, B., Li, M., Zhang, L.: From gene trees to species trees. SIAM J. Comput. **30**(3), 729–752 (2000)

25. Maddison, W.P., Maddison, D.: Mesquite: a modular system for evolutionary analysis (2015)

26. Page, R.: From gene to organismal phylogeny: reconciled trees and the gene tree/species tree problem. Mol. Phylogenet. Evol. **7**(2), 231–240 (1997)

27. Page, R.D.M.: Maps between trees and cladistic analysis of historical associations among genes, organisms, and areas. Syst. Biol. **43**(1), 58–77 (1994)

28. Park, H.J., Jin, G., Nakhleh, L.: Bootstrap-based support of hgt inferred by maximum parsimony. BMC Evol. Biol. **10**(1), 1–11 (2010)

29. Rambaut, A., Grassly, N.C.: Seq-Gen: an application for the Monte-Carlo simulation of DNA sequence evolution along phylogenetic trees. Comput. Appl. Biosci. **13**, 235–238 (1997)

30. Rasmussen, M.D., Kellis, M.: Unified modeling of gene duplication, loss, and coalescence using a locus tree. Genome Res. **22**(4), 755–765 (2012)

31. Sanderson, M.J., McMahon, M.M.: Inferring angiosperm phylogeny from EST data with widespread gene duplication. BMC Evol. Biol. **7**(Suppl. 1), S3 (2007)

32. Swenson, K., Doroftei, A., El-Mabrouk, N.: Gene tree correction for reconciliation and species tree inference. Algorithm Mol. Biol. **7**(1), 31 (2012)

33. The Génolevures Consortium: Génolevures: protein families and synteny among complete hemiascomycetous yeast proteomes and genomes. Nucleic Acids Res. **37**(Suppl. 1), D550–D554 (2009)

34. Wu, Y., Rasmussen, M.D., Bansal, M.S., Kellis, M.: Treefix: statistically informed gene tree error correction using species trees. Syst. Biol. **62**, 110–120 (2012)

35. Yu, Y., Warnow, T., Nakhleh, L.: Algorithms for MDC-based multi-locus phylogeny inference. In: Bafna, V., Sahinalp, S.C. (eds.) RECOMB 2011. LNCS, vol. 6577, pp. 531–545. Springer, Heidelberg (2011)
36. Zmasek, C., Eddy, S.: Rio: Analyzing proteomes by automated phylogenomics using resampled inference of orthologs. BMC Bioinform. **3**(1), 14 (2002)

Robustness of the Parsimonious Reconciliation Method in Cophylogeny

Laura Urbini[1,2,3], Blerina Sinaimeri[1,2,3(✉)], Catherine Matias[4],
and Marie-France Sagot[1,2,3]

[1] Université Lyon 1, Villeurbanne, France
[2] CNRS, UMR5558, Laboratoire de Biométrie Et Biologie Évolutive,
69622 Villeurbanne, France
[3] INRIA Grenoble Rhône - Alpes, Montbonnot-Saint-Martin, France
{laura.urbini,blerina.sinaimeri,Marie-france.sagot}@inria.fr
[4] Sorbonne Universités, Université Pierre et Marie Curie, Université Paris Diderot,
Centre National de la Recherche Scientifique, Laboratoire de Probabilités et Modèles
Aléatoires, 4 Place Jussieu, Paris, France
catherine.matias@math.cnrs.fr

Abstract. The aim of this paper is to explore the robustness of the parsimonious host-symbiont tree reconciliation method under editing or small perturbations of the input. The editing involves making different choices of unique symbiont mapping to a host in the case where multiple associations exist. This is made necessary by the fact that no tree reconciliation method is currently able to handle such associations. The analysis performed could however also address the problem of errors. The perturbations are re-rootings of the symbiont tree to deal with a possibly wrong placement of the root specially in the case of fast-evolving species. In order to do this robustness analysis, we introduce a simulation scheme specifically designed for the host-symbiont cophylogeny context, as well as a measure to compare sets of tree reconciliations, both of which are of interest by themselves.

Keywords: Cophylogeny · Parsimony · Event-based methods · Robustness · Measure for tree reconciliation comparison

1 Introduction

Almost every organism in the biosphere is involved in a so-called *symbiotic* interaction with other biological species, that is, in an interaction which is close and often long term. Such interactions (one speaks also of *symbiosis*) can involve two or more species and be of different types, ranging from mutualism (when both species benefit) to parasitism (when one benefits to the detriment of the other). Some interactions may even become obligatory in the sense that neither species is able anymore to live without the other. This may in particular be the case when one of the species lives inside the cells of the other. We speak then of *endosymbiosis* (notice however that not all endosymbioses are obligatory). Understanding symbiosis in general is therefore important in many different areas of biology.

© Springer International Publishing Switzerland 2016
M. Botón-Fernández et al. (Eds.): AlCoB 2016, LNBI 9702, pp. 119–130, 2016.
DOI: 10.1007/978-3-319-38827-4_10

As symbiotic interactions may continue over very long periods of time, the species involved can affect each other's evolution. This is known as *coevolution*. Studying the joint evolutionary history of species engaged in a symbiotic interaction enables in particular to better understand the long-term dynamics of such interactions. This is the subject of *cophylogeny*.

The currently most used method in cophylogenetic studies is the so-called *phylogenetic tree reconciliation* [3,4,12,16]. In this model, we are given the phylogenetic tree of the hosts H, the one of the symbionts S, and a mapping ϕ from the leaves of S to the leaves of H indicating the known symbiotic relationships among present-day organisms. In general, the common evolutionary history of the hosts and of their symbionts is explained through four main macro-evolutionary events that are assumed to be recovered by the tree reconciliation: (a) cospeciation, when host and symbiont speciate together; (b) duplication, when the symbiont speciates but not the host; (c) host switch, when after speciation of the symbiont, one of the new species of symbionts switches to a new host that is not related to the previous one; and (d) loss, which can describe three different and undistinguishable situations: (i) speciation of the host species independently of the symbiont, which then follows just one of the new host species due to factors such as, for instance, geographical isolation; (ii) cospeciation of host and symbiont, followed by extinction of one of the new symbiont species and; (iii) same as (ii) with failure to detect the symbiont in one of the two new host species. A reconciliation is a function λ which is an extension of the mapping ϕ between leaves to a mapping that includes all internal nodes and that can be constructed using the four types of events above. An optimal reconciliation is usually defined in a parsimonious way: a cost is associated to each event and a solution of minimum total cost is searched for. If timing information (*i.e.* the order in which the speciation events occurred in the host phylogeny) is not known, as is usually the case, the problem is NP-hard [15,23]. A way to deal with this is to allow for solutions that may be biologically unfeasible, that is for solutions where some of the switches induce a contradictory time ordering for the internal nodes of the host tree. In this case, the problem can be solved in polynomial time [1,6,7,13,21]. In most situations, as shown in [6], among the many optimal solutions, some are time-feasible.

However, an important issue in this model is that it makes strong assumptions on the input data which may not be verified in practice. We examine two cases where this situation happens.

– The first is related to a limitation in the currently available methods for tree reconciliation where the association ϕ of the leaves is for now, to the best of our knowledge, required to be a function. A leaf s of the symbiont tree can therefore be mapped to at most one leaf of the host tree. This is clearly not realistic as a single symbiont species can infect more than one host. We henceforth use the term *multiple association* to refer to this phenomenon. For each present-day symbiont involved in a multiple association, one is currently forced to choose a single one. Clearly, this may have an influence on the solutions obtained.

– The second case addresses a different type of problem related to the phylogenetic trees of hosts and symbionts. These indeed are assumed to be correct, which may not be the case already for the hosts even though these are in general eukaryotes for which relatively accurate trees can be inferred, and can become really problematic for the symbionts which most often are prokaryotes and can recombine among them [14,20,22]. We do not address the problem of recombination in this paper, but another one that may also have an influence in the tree reconciliation. This is the problem of correctly rooting a phylogenetic tree. Many phylogenetic tree reconstruction algorithms in fact produce unrooted trees [14,19,22]. The outgroup method is the most widely used in phylogenetic studies but a correct indication of the root position strongly depends on the availability of a proper outgroup [9,18,20]. A wrong rooting of the trees given as input may lead to an incorrect output.

The aim of this paper is, in the two cases, to explore the robustness of the parsimonious tree reconciliation method under "editing" (multiple associations) or "small perturbations" of the input (rooting problem). Notice that the first case is in general due to the fact that we are not able for now to handle multiple associations, although there could also be errors present in the association of the leaves that is given as input. The editing or perturbations we will be considering involve, respectively: (a) making different choices of single symbiont-host leaf mapping in the presence of multiple associations, and (b) re-rooting of the symbiont tree. In both studies, we explore the influence of six cost vectors that are commonly used in the literature (for a more detailed discussion, see for e.g. [2,4]). The final objective is to arrive at a better understanding of the relationship between the input and output of a parsimonious tree reconciliation method, and therefore at an evaluation of the confidence we can have in the output.

We wish here to call attention to the fact that we will consider the robustness of the parsimonious method in the case where the solutions provided may be time-unfeasible. Our choice is driven by two reasons. The first is that, as already mentioned, finding time-feasible optimal tree reconciliations is an NP-hard problem, and therefore testing a significant number of large datasets is computationally impossible in practice. The second is that, as also indicated, it has been empirically observed that time-unfeasible methods when they are exhaustive, that is when they correctly output all optimal solutions, can be a good heuristic for finding optimal time-feasible solutions [6]. Many tree reconciliation algorithms exist, but only a few enumerate all solutions. The most commonly used are NOTUNG [21], JANE 4 [5], and CORE-PA [13]. However, the first was designed for a gene/species context and imposes some restrictions on the costs that may be given to some of the events, while the last two provide for most instances only a proper subset of all the optimal solutions [6]. Currently, only the method that we developed, called EUCALYPT [6], is exhaustive, and we therefore decided to use it exclusively in order to explore the robustness of the parsimonious tree reconcilation method.

Another important point is that we tested the parsimonious reconciliation method both on real and simulated datasets. There are not many methods available to simulate datasets that coevolved as these were mostly developed in a gene/species context [1,7]. These are not suitable here for two reasons, the first being that they do not consider cospeciation as an event with its own parameter value (a gene *automatically* speciates within its species, *i.e.* when speciation occurs we consider that two different genes are automatically created, whether their sequences/functions already differ or not). The second reason is that these methods most often rely on a dating scheme of the host tree which might be difficult to tune so as to mimic real datasets. These limitations were already noticed in [10] where the authors attempted to provide their own simulation setup (to our knowledge, the only other one available in the cophylogeny context) by generating simultaneously a host and a symbiont tree relying on parameter values for the events. In this paper, we use a simulation method which we previously introduced in COALA [2] whose interest lies in that it uses parameter values (for the event probabilities) that are estimated on real datasets. Hence, this simulation scheme is more realistic and is designed for the cophylogeny context.

We start by introducing the datasets that will be used, both real and simulated ones as well as in the latter case, the method to generate them. We also present a measure to compare sets of tree reconciliations which may be of independent interest. We then describe the methods used to explore small perturbations in the two cases considered here, and discuss the results obtained.

The implemented methods are included in the tree reconciliation method we previously developed, called EUCALYPT, and will be made freely available at http://eucalypt.gforge.inria.fr/. This webpage also contains the online Supplementary Material with exhaustive results on the datasets.

2   Materials and Methods

In what follows, a dataset is a pair of host and symbiont trees (H, S), together with the association ϕ of their leaves. The indexes c, d, s, l relate to the 4 different events: cospeciation, duplication, switch and loss, respectively.

To analyse the influence of a perturbation, we adopted a set of cost events that correspond to those most commonly used in the literature on cophylogeny. We thus considered the following cost vectors $c = \langle c_c, c_d, c_s, c_l \rangle \in \mathcal{C}$ where $\mathcal{C} = \{\langle -1,1,1,1 \rangle, \langle 0,1,1,1 \rangle, \langle 0,1,2,1 \rangle, \langle 0,2,3,1 \rangle, \langle 1,1,1,1 \rangle, \langle 1,1,3,1 \rangle\}$.

2.1   Materials

Biological Datasets. To test the robustness of the method, we selected 15 biological datasets from the literature: AW - Arthropods (12 leaves) & *Wolbachia* (12 leaves), CT - *Cichlidogyrus* (19 leaves) & *Tropheini* (28 leaves), EC - *Encyrtidae* (7 leaves) & *Coccidae* (10 leaves), FD - Fishes (20 leaves) & *Dactylogyrus* (50 leaves), GL - Gophers (8 leaves) & Lices (10 leaves), IFL - Insects (17 leaves) & Flavobacterial endosymbionts (17 leaves), MP - *Myrmica* (8 leaves) & *Phengaris* (8 leaves), PML - Pelicans (18 leaves) & Lices (18 leaves) where both trees are generated through a maximum likelihood approach,

PMP - Pelicans (18 leaves) & Lices (18 leaves) where both trees are generated through a maximum parsimony approach, PP - Primates (36 leaves) & Pinworms (40 leaves), RH - Rodents (34 leaves) & Hantaviruses (42 leaves), RP- Rodents (13 leaves) & Pinworms (13 leaves), SBL - Seabirds (15 leaves) & Lices (8 leaves), SC - Seabirds (11 leaves) & Chewing Lices (14 leaves) and SCF - Smut Fungi (15 leaves) & Caryophillaceus plants (16 leaves). The choice was dictated by: (1) the availability of the data in public databases, and (2) the desire to cover for situations as widely different as possible in terms of the topology of the trees and the presence of multiple associations. For a more detailed description of these biological datasets, see the online Supplementary Material. We call attention here to the fact that only 3 of these datasets present multiple associations (namely MP, SBL, SFC) and are the ones used for studying the robustness of the method in the case of multiple associations.

Simulated Datasets. We generated simulated datasets using a method that we previously developed, called COALA [2], and the 15 biological datasets as follows.

For any such dataset, COALA first estimates the corresponding probability of each coevolutionary event (cospeciation, duplication, switch and loss) based on an approximate Bayesian computation approach. As we needed the datasets to be as realistic as possible, each time we ran COALA to obtain 50 vectors of probabilities $\gamma = \langle \gamma_c, \gamma_d, \gamma_s, \gamma_l \rangle$ that are in some sense a likely explanation of the observed data.

In a second step, we used these vectors and the symbiont tree generation algorithm in COALA (see Baudet *et al.* [2] for more details) to obtain, for each vector γ, a simulated symbiont tree S' whose evolution follows that of the host tree H. Each dataset (H, S, ϕ) and probability vector γ thus led to a simulated dataset (H, S', ϕ'). In total, we created $15 \times 50 = 750$ such datasets. For each of the 15 real datasets, we call the whole set of 50 simulated datasets (generated using the parameter estimates on the real dataset) by the name of the real dataset followed by *sim*, for instance AW-sim.

The simulated datasets will be used only for testing the rooting of the trees. Indeed, using simulated datasets in the multiple associations context would require a model that allows for such multiple associations by considering additional events. To the best of our knowledge, such a model does not exist yet. We therefore did not use such datasets to test the robustness of the associations.

2.2   Methods

Generating All the Optimal Solutions. We used EUCALYPT [6], which for a given dataset (H, S, ϕ) and vector $c = \langle c_c, c_d, c_s, c_l \rangle$ specifying the costs of the events, generates all the optimal reconciliations in polynomial-delay, meaning that the computation time between two outputs is polynomial in the input size.

Comparing Two Sets of Reconciliations. To estimate the similarity of the outputs of two different runs of the tree reconciliation algorithm, we needed a measure to compare two sets of tree reconciliations. Most studies summarise a reconciliation as a *pattern* of integers $\pi = \langle n_c, n_d, n_s, n_l \rangle$, representing the number of each event that it contains. The set of optimal solutions for a given dataset (H, S, ϕ) and cost vector c can thus be viewed as a multiset $\Lambda_{H,S,\phi,c}$ of patterns in $\mathbb{N}^4$. Notice that we needed to consider multisets as different reconciliations may induce the same pattern of events.

There is a wide literature on distances for sets of points. One of the best-known metrics between subsets, the Hausdorff metric, does not take into account the overall structure of the point sets. Other distances used for mining multisets, such as the Jaccard or Minkowski distance (see for example Chap. 6 in [11]), have the drawback of taking into account not the distance between the elements in the sets but only the number of different elements and their multiplicity.

Hence, for our purpose, we decided to introduce the following measure. Given a tree reconciliation Λ (i.e. a multiset of patterns), we define its representative $v_\Lambda = \sum_{\pi \in \Lambda} \pi$. Notice that such sum takes into account the multiplicities of a pattern. Given two tree reconciliations Λ_1 and Λ_2, we define a *dissimilarity measure* $d(\Lambda_1, \Lambda_2)$ as follows:

$$d(\Lambda_1, \Lambda_2) = \frac{||v_{\Lambda_1} - v_{\Lambda_2}||}{(|\Lambda_1| + |\Lambda_2|) \max_{\pi \in \Lambda_1 \cup \Lambda_2} ||\pi||} \tag{1}$$

where $|| \cdot ||$ is the L_1 norm and $|\Lambda|$ is the cardinality of the multiset Λ. Observe that $d(\Lambda_1, \Lambda_2) = 0$ whenever $\Lambda_1 = \Lambda_2$ while the converse is not necessarily true. Note also that we normalised this dissimilarity measure so that it takes values in $[0, 1]$. This dissimilarity measure, while not being a distance, enables us to summarize the comparison between two multisets of reconciliations. In particular, it takes into account both the multiplicity of the patterns and their actual values (patterns are vectors in $\mathbb{N}^4$ that might be close to each other).

Choosing Among Multiple Associations. Three of the real datasets we selected present multiple associations. For each of them, we considered all the datasets that may be obtained by resolving the multiple associations in all the possible ways. More precisely, for each symbiont associated with more than one host, we chose one and only one of the possible associations, and we did this in all the possible ways. For instance, in the SBL dataset, 5 out the 8 leaves of the symbiont tree have multiple associations, each connected to 2, 2, 4, 5, and 7 leaves of the host tree respectively (see Fig. 1 in the online Supplementary Material). By choosing in all possible ways among the multiple associations, we thus obtain 560 datasets.

Re-rooting of the Symbiont Tree. Most phylogenetic reconstruction algorithms produce unrooted trees, or rooted ones that have an unreliable root [9]. Rooting a phylogenetic tree is especially challenging for fast-evolving organisms. We therefore studied the influence on the optimal tree reconciliation of an erroneous rooting of the symbiont tree. More precisely, given a host tree H and a symbiont tree S, the association of their leaves ϕ, and a cost vector c, we compute all the optimal reconciliations for the pair H, S' where S' is obtained by positioning the root of S in an edge of S. Intuitively, one would expect that the correct positioning of the root would correspond to the reconciliation(s) having the minimum cost among all the ones that could be obtained by other rootings. This is indeed motivated by the same parsimony principle as for the tree reconciliation itself. Although slightly less immediate to grasp, one could expect also that positioning the root "near" to what would be the real one would lead to optimal reconciliation costs that are near the minimum.

Both cases were in fact observed by Gorecki *et al.* [8] who showed the existence of a certain property in models such as the Duplication-Loss for the gene/species tree reconciliation. Such property, which the authors called the *plateau property*, states that if we assign to each edge of the parasite tree a value indicating the cost of an optimal reconciliation when considering the parasite tree rooted in that edge, the edges with minimum value form a connected subtree in the parasite tree, hence the name of plateau. Furthermore, the edge values in any path from a plateau towards a leaf are monotonically increasing. In the presence of host switches, it was however not known whether such plateau property was satisfied.

Here, for both biological and simulated datasets, we count the number of plateaux (i.e. subtrees where rootings lead to minimal optimum cost), and we further keep track whether the original root belongs to a plateau. To study the robustness, we define a "small perturbation" of the rooting as follows. Given a dataset (H, S, ϕ), let $k = \max(5\,\%|V(S)|, 3)$. We compute all the optimal reconciliations for the pair H, S' where S' is obtained from S by positioning the root of S in an edge $(x, y) \in E(S)$ at a distance exactly k from the root, the latter being defined as the minimum distance between the node and the edge endpoints. The variable k captures the "closeness" of the new root to the original one. We compare the sets of reconciliations obtained with the true positioning of the root and with the positioning at distance k using our dissimilarity measure (1). We then analyse the variations of these dissimilarities with respect to the variation of the distance k.

3   Results and Discussions

For both the editing of host-symbiont associations and perturbations of the symbiont tree root, we present only part of the results obtained in our analysis (in terms of datasets and/or of cost vectors) for reasons of space. In every case, the choice of which results to show was dictated either by the most interesting case observed among all those explored for the purposes of a discussion of the effect of edits and small perturbations on a parsimonious tree reconciliation, or, in the case of the cost vectors, by the one(s) that are more commonly used in the literature. An exhaustive presentation appears in Supplementary Material. Here, time-unfeasible reconciliations have been filtered-out. For each result appearing in Supplementary Material, we specify whether this is the case or not.

3.1   Perturbation of the Present-Day Host-Symbiont Associations

We present here the results for the SBL dataset analysed with cost vector $\langle 0, 1, 1, 1 \rangle$. The TreeMap analysis of this dataset performed in [17] tried to maximise the number of cospeciations between hosts and symbionts but found out that sometimes host switches must be postulated to maximise cospeciation. Thus in some sense the choice of this cost vector is in accordance with the TreeMap philosophy. Our results for this dataset with the other cost vectors together with the two other datasets (MP and SFC) are presented in Sect. 2.1 from the online Supplementary Material.

Figure 1 (left) shows the optimal reconciliation costs obtained for the 560 datasets that were simulated from the SBL one by resolving the multiple associations in all the possible ways. We observe that when we change the associations, most often the optimum cost remains the same, namely 70 % of the datasets have the same cost (of 7). However, in many cases (30 %), changing association of the leaves results in a change of the optimum cost value (from 7 to a value in $\{6,8,9\}$).

To go further and analyse whether two datasets with same optimum cost have the same evolutionary history, we compared their sets of reconciliation patterns as described in Sect. 2.2. Figure 1 (right) shows the pairwise dissimilarities (see Eq. 1) between the reconciliation sets of the 392 datasets with same optimum cost of 7. Even if often the dissimilarity between two reconciliation sets is 0 (and we checked that the multisets of reconciliations are in fact exactly the same in those cases), in 65.5 % of the cases this is not so, and the value instead ranges inside [0.05,0.6], the largest dissimilarity (value of 0.6) being observed in 8.5 % of the cases.

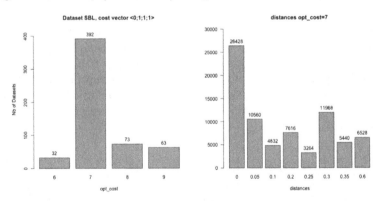

Fig. 1. Barplots of optimum cost (left) and dissimilarity between pairs of reconciliations with optimum cost 7 (right) obtained on the datasets derived from the SBL dataset by resolving the multiple associations in all the possible ways and computed with the cost vector $\langle 0, 1, 1, 1 \rangle$.

3.2   Re-rooting of the Symbiont Tree

Testing the Plateau Property. Table 1 in the online Supplementary Material presents the results for the 15 biological datasets evaluated with the 6 cost vectors in $\mathcal{C}$. Most of the datasets present only 1 plateau and only 2 datasets (CT and EC) present 2 plateaux. Moreover for 5 out of the 6 cost vectors tested, there is always a biological dataset for which 2 plateaux are observed.

The plateau property therefore does not hold in the presence of host switches for real datasets analysed with biologically plausible setups. It is interesting to observe that among the 15 biological datasets, there were never more than 2 plateaux. This may be due to the relatively small size of the trees.

We also note that in 37 % of the cases, the original root is not in a plateau. Moreover, the difference between the optimal cost obtained for the original rooting and the cost obtained by placing the root inside the plateau is quite large (difference between columns D and B in Table 1 in online Supplementary Material). Among these 37 %, in addition, for the datasets AW, FD, RH, and SFC, the original root of the symbiont tree is never in a plateau. This may indicate that either the original root is not at its correct position, or that there is not enough evolutionary dependence between the two organisms to allow for a correct inference of the symbiont tree root.

The simulated datasets present similar results as the biological ones (Table 2 in the online Supplementary Material). The number of datasets with more than one plateau however increases, as does in some cases the number of plateaux observed. Indeed, some simulated datasets from the sets AW-sim, MP-sim, and SFC-sim exhibit up to 5 plateaux. In 17 % of the simulations, the original root does not belong to a plateau (data not shown).

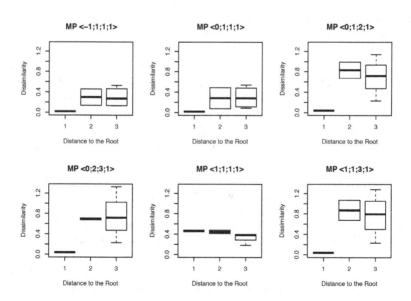

Fig. 2. Boxplots of the dissimilarities between reconciliations obtained for the original dataset MP and all datasets simulated from MP by re-rooting the symbiont tree at distance k from the original root. The six plots correspond to the 6 cost vectors in $\mathcal{C}$. The x-axis shows the distance k between new and original root. The y-axis shows the value d of the dissimilarity of the reconciliation patterns.

Rerooting at Distance k. We show in Fig. 2 the results obtained with the biological dataset MP. Similar figures are presented with other biological datasets in Sect. 2.3 from the online Supplementary Material. Here the dissimilarity of the reconciliation globally increases as k also increases. The farther is the new root from the original one, the more dispersed the patterns tend to be (*i.e.* the values of d have larger variance). These conclusions extend for 8 of the remaining biological datasets (EC, FD, GL, PML, PP, RP, SBL, SC). However, no such global trend is obtained for the other biological datasets for which we only observe variability (neither increasing nor decreasing) in the dissimilarities.

As concerns the simulated datasets, we observe a bigger dispersion between patterns with larger values taken by the dissimilarities (see Sect. 2.4 from the online Supplementary Material). This might be due to the fact that there are much more datasets (50 simulated datasets corresponding to one biological dataset). The trend of a global increase of the values and the variance of the dissimilarity when k increases is observed again.

4   Conclusions and Open Problems

In this paper, we explored the robustness of the parsimonious tree reconciliation method to some editing of the input required in order to associate a symbiont to a unique host in the case where multiple associations exist, as well as to small perturbations linked to a re-rooting of the symbiont tree.

In the first case, we observed that the choice of leaf associations may have a strong impact on the variability of the reconciliation output. Although such impact appears not so important on the cost of the optimum solution, probably due to the relatively small size of the input trees, the difference becomes more consequent when we refine the analysis by comparing, not the overall cost, but instead the patterns observed in the optimal solutions. Notice that this highlights the great interest in finding measures for the dissimilarity of sets of reconciliations such as the new one we proposed in this paper.

As indicated, we were able to do the analysis on the choice of leaf associations only for the real biological datasets because we are currently not capable of simulating the coevolution of symbionts and hosts following the phylogenetic tree of the latter and allowing for an association of the symbionts to multiple hosts. This is an interesting and we believe important open problem in the literature on reconciliations which we are currently trying to address.

As concerns the problem of the rooting, we were able to show that allowing for host switches invalidates the plateau property that had been previously observed (and actually also mathematically proved) in the cases where such events were not considered. Again here, the number of plateaux observed is small for the real datasets (this number is indeed 2). Moreover, such increase from 1 to 2 does not concern all pairs of datasets and of cost vectors, even though for all, except one of the cost vectors tested, there is always a biological dataset for which 2 plateaux are observed. We might be tempted to say that this is once more due to the small sizes of the input trees. However, the sizes are of the same order for the simulated datasets, but there the differences are greater: we may indeed reach up to 5 plateaux in some cases. We are currently not able to explain this difference between the two types of datasets (this might be just chance related to the fact that we have 50 times more simulated than biological datasets). For both of them, we also observe that the original root may not be inside a plateau, and that the proportion for which this is observed is approximately the same (3 cases out of 15 as compared to 17 % respectively) for real or simulated datasets. We hypothesised that for the real datasets, this might indicate that the original root is not at its correct position. It would be interesting in future to try to validate this hypothesis. If it were proved to be true, an interesting, but hard open problem would be to be able to use as input for a cophylogeny study unrooted trees instead of rooted one, or even directly the sequences that were originally used to infer the host and symbiont trees. In this case, we would then have to, at a same time, infer the trees and their optimal reconciliation.

Re-rooting the symbiont tree at distance k leads in many cases to an increase in both the values and variance of the dissimilarity measure in the patterns (9 out of 15 biological datasets and all sets of simulations). The dispersion and the values of dissimilarity are also greater in the simulated datasets than in the biological ones (here again, this could be an artefact due to the large number of simulated datasets).

Clearly, the effect in terms of number of plateaux depends on the presence of host switches since this number was proved to be always one when switches are not allowed [8]. Perhaps the most interesting open problem now is whether there is a relation between the number of plateaux observed as well as the level of dissimilarity among the patterns obtained on one hand, and the number of host switches in the optimal solutions on the other hand. Actually the relation may be more subtle, and be related not to the number of switches but to the distance involved in a switch, where by distance of a switch we mean the evolutionary distance between the two hosts involved in it. This could be measured in terms of the number of branches (as is the case in our method EUCALYPT) or in terms of the sum of the branch lengths, that is of estimated evolutionary time.

Acknowledgments. The authors would like to express their gratitude to Christian Gautier for fruitful preliminary discussions on this work.

References

1. Bansal, M.S., Alm, E.J., Kellis, M.: Efficient algorithms for the reconciliation problem with gene duplication, horizontal transfer and loss. Bioinf. **28**(12), i283–i291 (2012)
2. Baudet, C., Donati, B., Sinaimeri, B., Crescenzi, P., Gautier, C., Matias, C., Sagot, M.F.: Cophylogeny reconstruction via an approximate Bayesian computation. Syst. Biol. **64**(3), 416–431 (2015)
3. Charleston, M.A.: Jungles: a new solution to the host/parasite phylogeny reconciliation problem. Math. Biosci. **149**(2), 191–223 (1998)
4. Charleston, M.A.: Recent results in cophylogeny mapping. Adv. Parasitol. **54**, 303–330 (2003)

5. Conow, C., Fielder, D., Ovadia, Y., Libeskind-Hadas, R.: Jane: a new tool for the cophylogeny reconstruction problem. Algo. Mol. Biol. **5**(16), 1–10 (2010)
6. Donati, B., Baudet, C., Sinaimeri, B., Crescenzi, P., Sagot, M.F.: EUCALYPT: efficient tree reconciliation enumerator. Algo. Mol. Biol. **10**(1), 3 (2014)
7. Doyon, J.-P., Scornavacca, C., Gorbunov, K.Y., Szöllősi, G.J., Ranwez, V., Berry, V.: An efficient algorithm for gene/species trees parsimonious reconciliation with losses, duplications and transfers. In: Tannier, E. (ed.) RECOMB-CG 2010. LNCS, vol. 6398, pp. 93–108. Springer, Heidelberg (2010)
8. Górecki, P., Eulenstein, O., Tiuryn, J.: Unrooted tree reconciliation: a unified approach. IEEE/ACM Trans. Comput. Biology Bioinf. **10**(2), 522–536 (2013)
9. Holland, B., Penny, D., Hendy, M.: Outgroup misplacement and phylogenetic inaccuracy under a molecular clock: a simulation study. Syst. Biol. **52**(2), 229–238 (2003)
10. Keller-Schmidt, S., Wieseke, N., Klemm, K., Middendorf, M.: Evaluation of host parasite reconciliation methods using a new approach for cophylogeny generation. Technical report, University of Leipzig (2011). https://www.bioinf.uni-leipzig.de/Publications/PREPRINTS/11-013.pdf
11. Kosters, W.A., Laros, J.F.J.: Metrics for mining multisets. In: Bramer, M., Coenen, F., Petridis, M. (eds.) Research and Development in Intelligent Systems XXIV, pp. 293–303. Springer, London (2008)
12. Merkle, D., Middendorf, M.: Reconstruction of the cophylogenetic history of related phylogenetic trees with divergence timing information. Theory Biosci. **123**(4), 277–299 (2005)
13. Merkle, D., Middendorf, M., Wieseke, N.: A parameter-adaptive dynamic programming approach for inferring cophylogenies. BMC Bioinf. **11**(Suppl 1), S60 (2010)
14. Nei, M., Kumar, S.: Molecular evolution and phylogenetics. Oxford Univ, Press (2000)
15. Ovadia, Y., Fielder, D., Conow, C., Libeskind-Hadas, R.: The cophylogeny reconstruction problem is NP-complete. J. Comput. Biol. **18**(1), 59–65 (2011)
16. Page, R.D.M.: Parallel phylogenies: reconstructing the history of host-parasite assemblages. Cladistics **10**(2), 155–173 (1994)
17. Paterson, A.M., Gray, R.D., Clayton, D.H., Moore, J.: Host-parasite co-speciation, host switching, and missing the boat. In: Clayton, D.H., Moore, J. (eds.) Host-parasite evolution: general principles and avian models, pp. 236–250. Oxford University Press, Oxford (1997)
18. Qiu, Y.L., Lee, J., Whitlock, B.A., Bernasconi-Quadroni, F., Dombrovska, O.: Was the anita rooting of the angiosperm phylogeny affected by long-branch attraction? Mol. Biol. Evol. **18**(9), 1745–1753 (2001)
19. Sanderson, M.J., Shaffer, H.B.: Troubleshooting molecular phylogenetic analyses. Annu. Rev. Ecol. Syst. **33**, 49–72 (2002)
20. Stavrinides, J., Guttman, D.S.: Mosaic evolution of the severe acute respiratory syndrome coronavirus. J. Virol. **78**(1), 76–82 (2004)
21. Stolzer, M., Lai, H., Xu, M., Sathaye, D., Vernot, B., Durand, D.: Inferring duplications, losses, transfers and incomplete lineage sorting with nonbinary species trees. Bioinf. **28**(18), i409–i415 (2012)
22. Swofford, D.L., Olsen, G.J., Waddell, P.J., Hillis, D.M.: Phylogenetic inference. In: Hillis, D.M., Moritz, C., Mable, B.K. (eds.) Molecular systematics, pp. 407–514. Sinauer Associates Inc, Sunderland (1996)
23. Tofigh, A., Hallett, M., Lagergren, J.: Simultaneous identification of duplications and lateral gene transfers. IEEE/ACM Trans. Comput. Biol. Bioinf. (TCBB) **8**(2), 517–535 (2011)

Sequence Analysis and Rearrangement

Sorting with Forbidden Intermediates

Carlo Comin[1,2($\boxtimes$)], Anthony Labarre[2], Romeo Rizzi[3], and Stéphane Vialette[2]

[1] Department of Mathematics, University of Trento, Trento, Italy
Carlo.Comin@unitn.it
[2] Université Paris-Est, LIGM (UMR 8049), UPEM, CNRS, ESIEE, ENPC,
77454 Marne-la-Vallée, France
{Anthony.Labarre,Stephane.Vialette}@u-pem.fr
[3] Department of Computer Science, University of Verona, Verona, Italy
Romeo.Rizzi@univr.it

Abstract. A wide range of applications, most notably in comparative genomics, involve the computation of a shortest sorting sequence of operations for a given permutation, where the set of allowed operations is fixed beforehand. Such sequences are useful for instance when reconstructing potential scenarios of evolution between species, or when trying to assess their similarity. We revisit those problems by adding a new constraint on the sequences to be computed: they must *avoid* a given set of *forbidden intermediates*, which correspond to species that cannot exist because the mutations that would be involved in their creation are lethal. We initiate this study by focusing on the case where the only mutations that can occur are exchanges of any two elements in the permutations, and give a polynomial time algorithm for solving that problem when the permutation to sort is an involution.

Keywords: Genome rearrangement · Permutation sorting · Lethal mutations · Forbidden vertices · Hypercube graphs · st-Connectivity

1 Introduction

Computing distances between permutations, or sequences of operations that transform them into one another, are two generic problems that arise in a wide range of applications, including comparative genomics [7], ranking [5], and interconnection network design [15]. Those problems are well-known to reduce to constrained sorting problems of the following form: given a permutation π and a set S of allowed operations, find a sequence of elements from S that sorts π and is as short as possible. In the context of comparative genomics, the sequence to be reconstructed yields a possible scenario of evolution between the genomes represented by π and the target identity permutation ι, where all permutations obtained in-between are successive descendants of π (and ancestors of ι). The many possible choices that exist for S, as well as other constraints or cost functions with which they can be combined, have given rise to a tremendous number of variants whose algorithmic and mathematical aspects have now been studied for decades [7]. Specific issues that biologists feel need to be addressed to

© Springer International Publishing Switzerland 2016
M. Botón-Fernández et al. (Eds.): AlCoB 2016, LNBI 9702, pp. 133–144, 2016.
DOI: 10.1007/978-3-319-38827-4_11

improve the applicability of these results in a biological context include: (1) the over simplicity of the model (permutations do not take duplications into account), (2) the rigid definition of allowed operations, which fails to capture the complexity of evolution, and (3) the complexity of the resulting problems, where algorithmic hardness results abound even for deceivingly simple problems. A large body of work has been devoted to addressing those issues, namely by proposing richer models for genomes, encompassing several operations with different weights [7]. Some approaches for increasing the reliability of rearrangement methods by adding additional biologically motivated constraints have been investigated (see for example [2] for conserved intervals, [8] for restricting the set of allowed inversions and [1] for preserving the number of inversions in the scenario which commute with all common intervals). However, another critical issue has apparently been overlooked: to the best of our knowledge, no model takes into account the fact that the solutions it produces may involve allele mutations that are lethal to the organism on which they act. *Lethals* are usually a result of mutations in genes that are essential to growth or development [10]; they have been known to occur for more than a century [4], as they were first discovered by Cuénot in 1905 while studying the inheritance of coat colour in mice. As a consequence, solutions that may be perfectly valid from a mathematical point of view should nonetheless be rejected on the grounds that some of the intermediate ancestors they produce are nonviable and can therefore not have had any descendants. We revisit the family of problems mentioned above by adding a natural constraint which, as far as we know, has not been previously considered in this form (see e.g. [1,2,8] for connected attempts): namely, the presence of a set of forbidden intermediate permutations, which the sorting sequence that we seek must avoid. We refer to this family of problems as GUIDED SORTING problems, since they take additional guidance into account. In this paper we focus our study on the case where only *exchanges* (i.e., algebraic transpositions) are allowed; furthermore, we simplify the problem by demanding that the solutions we seek be *optimal* in the sense that no shorter sorting sequence of exchanges exists even when no intermediate permutation is forbidden. We choose to focus on exchanges because of their connection to the underlying *disjoint cycle structure* of permutations, which plays an important role in many related sorting problems where a similar cycle-based approach, using this time the ubiquitous *breakpoint graph*, has proved extremely fruitful [14]. Therefore, we believe that progress on this particular variant will be helpful when attempting to solve related variants based on more complex operations. Our main contribution in this work is a polynomial time algorithm for solving GUIDED SORTING by exchanges when the permutation to sort is an *involution*. We show that, in that specific case, the space of all feasible sorting sequences admits a suitable description in terms of directed (s,t)-paths in hypercube graphs. We achieve this result by reducing GUIDED SORTING to the problem of finding directed (s,t)-paths that avoid a prescribed set $\mathcal{F} \subseteq V$ of *forbidden vertices*. Our main contribution, therefore, consists in solving this latter problem in time polynomial in just the encoding of $\mathcal{F}$ if G is constrained to be a *hypercube* graph, which is a novel algorithmic result that may be of independent interest. Specific properties that will be described later on [11,16] allow us to avoid the full construction of that graph, which would lead to an exponential time algorithm. We should mention that constrained variants of the (s,t)-connectivity problem have been studied already to some extent. For instance, already in the '70s, motivated by some problems in the field of automatic software testing and validation, [13] introduced the *path avoiding forbidden pairs* problem, namely, that of finding a directed (s,t)-path in a graph $G = (V, E)$ that contains at most one vertex from each pair in a prescribed set $\mathcal{F} \subseteq V \times V$ of *forbidden pairs* of vertices. [9] proved that the problem is NP-complete on DAGs. A number of special cases were shown to admit polynomial time algorithms, e.g. [18] studied the problem in directed graphs under a *skew-symmetry* condition. However, the involved techniques and the related results do not extend to our problem, for which we are aware of no previously known algorithm that runs in time polynomial in just the encoding of $\mathcal{F}$.

2   Background and Notation

Our aim is to sort a given permutation π using a predefined set of allowed operations, specified as a generating set S of the symmetric group $\mathfrak{S}_n$. We seek a sorting sequence that uses only elements from S and: (1) *avoids* a given set $\mathcal{F}$ of *forbidden permutations*, i.e. no intermediary permutation produced by applying the operations specified by the sorting sequence belongs to $\mathcal{F}$, and (2) is *optimal*, i.e. no shorter sorting sequence exists for π even if $\mathcal{F} = \emptyset$. We refer to the general problem of finding a sorting sequence under these constraints as GUIDED SORTING, and restrict in this paper the allowed operations to *exchanges* of any two elements (i.e. *algebraic transpositions*). For instance, let $\pi = \langle 2\ 3\ 1\ 4 \rangle$ and $\mathcal{F} = \{\langle 1\ 3\ 2\ 4 \rangle, \langle 3\ 2\ 1\ 4 \rangle\}$. Then $\langle 2\ 3\ 1\ 4 \rangle \mapsto \langle 2\ 1\ 3\ 4 \rangle \mapsto \langle 1\ 2\ 3\ 4 \rangle$ is a valid solution since it is optimal and avoids $\mathcal{F}$, but neither $\langle 2\ 3\ 1\ 4 \rangle \mapsto \langle 4\ 3\ 1\ 2 \rangle \mapsto \langle 4\ 3\ 2\ 1 \rangle \mapsto \langle 4\ 2\ 3\ 1 \rangle \mapsto \langle 1\ 2\ 3\ 4 \rangle$ nor $\langle 2\ 3\ 1\ 4 \rangle \mapsto \langle 1\ 3\ 2\ 4 \rangle \mapsto \langle 1\ 2\ 3\ 4 \rangle$ can be accepted: the former is too long, and the latter does not avoid $\mathcal{F}$.

We use standard notions and notation from graph theory (see e.g. [6] for undefined concepts), using $\{u, v\}$ (resp. (u, v)) to denote the edge (resp. arc) between vertices u and v of an undirected (resp. directed) graph $G = (V, E)$. All graphs we consider are *simple*: they contain neither loops nor parallel edges. If $\mathcal{F} \subseteq V$, a directed path $\mathsf{p} = v_0 v_1 \cdots v_n$ *avoids* $\mathcal{F}$ when $v_i \notin \mathcal{F}$ for every i. If $\mathcal{S} \subseteq V$ and $\mathcal{T} \subseteq V$, we say that a directed path p *goes from* $\mathcal{S}$ *to* $\mathcal{T}$ *in* G when p starts from some s in $\mathcal{S}$ and ends at some t in $\mathcal{T}$. When G is directed, we partition the neighbourhood $N(u)$ of a vertex u into the sets $N^{\text{out}}(u) = \{v \in V \mid (u, v) \in E\}$ and $N^{\text{in}}(u) = \{v \in V \mid (v, u) \in E\}$. Some of our graphs may be vertex-labelled, using any injective mapping $\ell : V \to \mathbb{N}$. For any $n \in \mathbb{N}$, $\wp_n = \wp([n])$ denotes the power set of $[n]$. The *hypercube graph on ground set* $[n]$, denoted by $\mathcal{H}_n$, is the graph with vertex set $\wp_n$ and in which the arc (U, V) connects vertices $U, V \subseteq [n]$ if there exists some $q \in [n]$ such that $U = V \setminus \{q\}$. If $S, T \in \wp_n$ and $|S| \leq |T|$, then $d_{S,T} = |T| - |S|$ is the *distance* between S and T. Finally, $\mathcal{H}_n^{(i)}$ denotes the family of all subsets of $\wp_n$ of size i.

3   Solving GUIDED SORTING For Involutions

The *Cayley graph* $\Gamma(\mathfrak{S}_n, S)$ of $\mathfrak{S}_n$ for a given generating set S of $\mathfrak{S}_n$ contains a vertex for each permutation in $\mathfrak{S}_n$ and an edge between any two permutations that can be obtained from one another using one element from S. A naïve approach for solving any variant of the GUIDED SORTING problem would build the part of $\Gamma(\mathfrak{S}_n, S)$ that is needed (i.e. without the elements of $\mathcal{F}$), then run a shortest path algorithm to compute an optimal sequence that avoids all elements of $\mathcal{F}$. This is highly impractical, since the size of Γ is exponential in n.

We describe in this section a polynomial time algorithm in the case of exchanges if π is an *involution*, i.e. a permutation such that for each $1 \leq i \leq n$, either $\pi_i = i$ or there exists an index j such that $\pi_i = j$ and $\pi_j = i$. From our point of view, involutions reduce to collections of disjoint pairs of elements that each need to be swapped by an exchange until we obtain the identity permutation, and the only forbidden permutations that could be produced by an optimal sorting sequence are involutions whose pairs of unsorted elements all appear in π. Therefore, we can reformulate our GUIDED SORTING problem in that setting as that of finding a directed (π, ι)-path in $\mathcal{H}_n$ that avoids all vertices in $\mathcal{F}$, where the permutation to sort π corresponds to the bottom vertex $\emptyset$ of $\mathcal{H}_n$ and the identity permutation ι corresponds to the top vertex $[n]$ of $\mathcal{H}_n$.

We shall focus on the following problem from here on.

PROBLEM: HY-STCON.

INPUT: the size $n \in \mathbb{N}$ of the underlying ground set $[n]$, a family of *forbidden vertices* $\mathcal{F} \subseteq \wp_n$, a *source* set $S \in \wp_n$ and a *target* set $T \in \wp_n$.
DECISION-TASK: Decide whether there exists a directed path p in $\mathcal{H}_n$ that goes from source S to target T avoiding $\mathcal{F}$;
SEARCH-TASK: Compute a directed path p in $\mathcal{H}_n$ that goes from source S to target T avoiding $\mathcal{F}$, provided that at least one such path exists.

We will show how to solve HY-STCON in time polynomial in $|\mathcal{F}|$ and n. The algorithm mainly consists in the continuous iteration of two phases:

1. *Double-BFS.* This phase explores the outgoing neighbourhood of the source S by a breadth-first search denoted by BFS$_\uparrow$ going from lower to higher levels of $\mathcal{H}_n$ while avoiding the vertices in $\mathcal{F}$. BFS$_\uparrow$ collects a certain (polynomially bounded) amount of visited vertices. Symmetrically, the incoming neighbourhood of the target vertex T is also explored by another breadth-first search BFS$_\downarrow$ going from higher to lower levels of $\mathcal{H}_n$ while avoiding the vertices in $\mathcal{F}$, also collecting a certain (polynomially bounded) amount of visited vertices.

2. *Compression.* If a valid solution has not yet been determined, then a compression technique is devised in order to shrink the size of the remaining search space. This is possible thanks to some nice regularities of the search space and to certain connectivity properties of hypercube graphs [11,16]. This allows us to reduce the search space in a suitable way and, therefore, to continue with the Double-BFS phase in order to keep the search towards valid solutions going.

Our main contribution is summarized in the following theorem. We devote the rest of this section to an in-depth description of the algorithms it mentions[1].

Theorem 1. *Concerning the* HY-STCON *problem, the following propositions hold on any input* $\langle S, T, \mathcal{F}, n \rangle$, *where* $d_{S,T}$ *is the distance between* S *and* T.

1. *There exists an algorithm for solving the* DECISION-TASK *of* HY-STCON *within* $O(\min(\sqrt{|\mathcal{F}| \, d_{S,T} \, n}, |\mathcal{F}|) \, |\mathcal{F}|^2 \, d_{S,T}^4 \, n^2)$ *time.*
2. *There exists an algorithm for solving the* SEARCH-TASK *of* HY-STCON *within* $O(\min(\sqrt{|\mathcal{F}| \, d_{S,T} \, n}, |\mathcal{F}|) \, |\mathcal{F}|^2 \, d_{S,T}^4 \, n^2 + |\mathcal{F}|^{5/2} n^{3/2} d_{S,T})$ *time.*

3.1   On Vertex-Disjoint Paths in Hypercube Graphs

The proof of Theorem 1 relies on connectivity properties of hypercube graphs [11]. The next result, which proves the existence of a family of certain vertex-disjoint paths in $\mathcal{H}_n$ that are called *Lehman-Ron paths*, will be particularly useful.

Theorem 2 (Lehman, Ron [16]). *Given* $n, m \in \mathbb{N}$, *let* $\mathcal{R} \subseteq \mathcal{H}_n^{(r)}$ *and* $\mathcal{S} \subseteq \mathcal{H}_n^{(s)}$ *with* $|\mathcal{R}| = |\mathcal{S}| = m$ *and* $0 \le r < s \le n$. *Assume there exists a bijection* $\varphi : \mathcal{S} \to \mathcal{R}$ *such that* $\varphi(S) \subset S$ *for every* $S \in \mathcal{S}$. *Then there exist* m *vertex-disjoint directed paths in* $\mathcal{H}_n$ *whose union contains all the subsets in* $\mathcal{S}$ *and* $\mathcal{R}$.

We call tuples $\langle \mathcal{R}, \mathcal{S}, \varphi, n \rangle$ that satisfy the hypotheses of Theorem 2 *Lehman-Ron tuples*, and we refer to the quantity $d = s - r$ as the *distance* between $\mathcal{R} \subseteq \mathcal{H}_n^{(r)}$ and $\mathcal{S} \subseteq \mathcal{H}_n^{(s)}$. An elementary inductive proof of Theorem 2 is in [16]. A careful and in-depth analysis of their proof, from the algorithmic perspective, yields a polynomial time algorithm for computing all the Lehman-Ron paths.

Theorem 3. *There exists an algorithm for computing all Lehman-Ron paths within time* $O(m^{5/2} n^{3/2} d)$ *on any Lehman-Ron input* $\langle \mathcal{R}, \mathcal{S}, \varphi, n \rangle$ *with* $|\mathcal{R}| = |\mathcal{S}| = m$, *where* d *is the distance between* $\mathcal{R}$ *and* $\mathcal{S}$ *and* n *is the size of the underlying ground set.*

In an extended version (i.e. [3]) we provide all the details of the above mentioned algorithm as well as a proof of the time complexity stated in Theorem 3, in which Menger's vertex-connectivity theorem [6] and Hopcroft-Karp's algorithm [12] for maximum cardinality matching in *undirected* bipartite graphs play a major role.

3.2   A Polynomial Time Algorithm for Solving HY-STCON

We now describe a polynomial time algorithm for solving HY-STCON, called `solve_HY-STCON()`, which takes as input an instance $\langle S, T, \mathcal{F}, n \rangle$ of HY-STCON, and returns a pair $\langle \text{YES}, \mathsf{p} \rangle$ where p is a directed path in $\mathcal{H}_n$ that goes from source S to target T avoiding $\mathcal{F}$ if such a path exists (otherwise, the algorithm

---

[1] See [3], Appendix B2 for correctness and Appendix B3 for complexity.

Algorithm 1. solving the HY-STCON problem.

Procedure $solve\_$HY-STCON$(S, T, \mathcal{F}, n)$

> **Input**: an instance $\langle S, T, \mathcal{F}, n \rangle$ of HY-STCON.
> **Output**: a pair $\langle \text{YES}, \text{p} \rangle$ where the path p is a solution to HY-STCON if such
> a path exists, NO otherwise.

1 $d_{S,T} \leftarrow |T| - |S|$;   // let $d_{S,T}$ be the distance between S and T
2 $\mathcal{S} \leftarrow \{S\}$; $\ell_\uparrow \leftarrow 0$;   // initialize the *frontier* $\mathcal{S}$ and its *level counter* $\ell_\uparrow$
3 $\mathcal{T} \leftarrow \{T\}$; $\ell_\downarrow \leftarrow 0$;   // initialize the *frontier* $\mathcal{T}$ and its *level counter* $\ell_\downarrow$
4 **while** *TRUE* **do**
5 | $\langle \mathcal{S}, \mathcal{T}, \ell_\uparrow, \ell_\downarrow \rangle \leftarrow$ double-bfs_phase$(\mathcal{S}, \mathcal{T}, \mathcal{F}, \ell_\uparrow, \ell_\downarrow, d_{S,T}, n)$;
6 | **if** $\mathcal{S} = \emptyset$ **OR** $\mathcal{T} = \emptyset$ **OR** $(\ell_\uparrow + \ell_\downarrow = d_{S,T}$ **AND** $\mathcal{S} \cap \mathcal{T} = \emptyset)$ **then**
7 | | **return** *NO*;
8 | **if** $\ell_\uparrow + \ell_\downarrow = d_{S,T}$ **AND** $\mathcal{S} \cap \mathcal{T} \neq \emptyset$ **then**
9 | | p $\leftarrow$ reconstruct_path$(\mathcal{S}, \mathcal{T}, n)$;
10 | | **return** $\langle \text{YES}, \text{p} \rangle$;
11 | returned_val $\leftarrow$ compression_phase$(\mathcal{S}, \mathcal{T}, \mathcal{F}, \ell_\uparrow, \ell_\downarrow, d_{S,T}, n)$;
12 | **if** $returned\_val = \langle \text{YES}, \text{p} \rangle$ **then return** p;
13 | **else** $\mathcal{T} \leftarrow$ returned_val;

---

simply returns NO). Algorithm 1 shows the pseudo-code for that procedure. The rationale at the base of solve_HY-STCON() consists in the continuous iteration of two major phases: double-bfs_phase() (line 5) and compression_ phase() (line 11). Throughout computation, both phases alternate repeatedly until a final state of termination is eventually reached (either at line 7, line 10 or line 12). At that point, the algorithm either returns a pair $\langle \text{YES}, \text{p} \rangle$ where p is the sought directed path, or a negative response NO instead[2]. We now describe both phases in more detail, and give the corresponding pseudo-code.

Breadth-First Search Phases. The first search BFS$_\uparrow$ starts from the source vertex S and moves upward, from lower to higher levels of $\mathcal{H}_n$. Meanwhile, it collects a certain (polynomially bounded) amount of vertices that do not lie in $\mathcal{F}$. In particular, at the end of any BFS$_\uparrow$ phase, the number of collected vertices will always lie between $|\mathcal{F}| \, d_{S,T} + 1$ and $|\mathcal{F}| \, d_{S,T} \, n$ (see line 1 of bfs_phase()). The set $\mathcal{S}$ of vertices collected at the end of BFS$_\uparrow$ is called the *(source) frontier* of BFS$_\uparrow$. All vertices within $\mathcal{S}$ have the same cardinality, i.e. $|X_1| = |X_2|$ for every $X_1, X_2 \in \mathcal{S}$. Also, the procedure keeps track of the highest level of depth $\ell_\uparrow$ that is reached during BFS$_\uparrow$. Thus, $\ell_\uparrow$ corresponds to the distance between the source vertex S and the current frontier $\mathcal{S}$, formally, $\ell_\uparrow = |X| - |S|$ for every $X \in \mathcal{S}$. Since at the beginning of the computation BFS$_\uparrow$ starts from the source vertex S, solve_HY-STCON() initializes $\mathcal{S}$ to $\{S\}$ and $\ell_\uparrow$ to 0 at line 2.

Similarly, the second search BFS$_\downarrow$ starts from the target vertex T and moves downward, from higher to lower levels of $\mathcal{H}_n$, also collecting a certain (polynomially bounded) amount of vertices that do not lie in $\mathcal{F}$. As in the previous case,

---

[2] See [3], Appendix B2-B3 for proofs of correctness and complexity.

Algorithm 2. Breadth-First-Search phases of solve_Hy-STCON().

Procedure $double\text{-}bfs\text{-}phase(\mathcal{S}, \mathcal{T}, \mathcal{F}, \ell_\uparrow, \ell_\downarrow, d_{S,T}, n)$

1    $\langle \mathcal{S}^*, \ell_\uparrow^* \rangle \leftarrow$ bfs_phase$(\mathcal{S}, \mathcal{F}, \ell_\uparrow, \ell_\downarrow, \text{out}, d_{S,T}, n)$;   // BFS$_\uparrow$ phase

2    $\langle \mathcal{T}^*, \ell_\downarrow^* \rangle \leftarrow$ bfs_phase$(\mathcal{T}, \mathcal{F}, \ell_\downarrow, \ell_\uparrow^*, \text{in}, d_{S,T}, n)$;   // BFS$_\downarrow$ phase

3   **return** $\langle \mathcal{S}^*, \mathcal{T}^*, \ell_\uparrow^*, \ell_\downarrow^* \rangle$;

SubProcedure $bfs\text{-}phase(\mathcal{X}, \mathcal{F}, \ell_x, \ell_y, \text{drt}, d_{S,T}, n)$

1   **while** $1 \le |\mathcal{X}| \le |\mathcal{F}| d_{S,T}$ **AND** $\ell_x + \ell_y < d_{S,T}$ **do**

2      $\mathcal{X} \leftarrow$ next_step_bfs$(\mathcal{X}, \mathcal{F}, \text{drt}, n)$;

3      $\ell_x \leftarrow \ell_x + 1$;

4   **return** $\langle \mathcal{X}, \ell_x \rangle$;

SubProcedure $next\text{-}step\text{-}bfs(\mathcal{X}, \mathcal{F}, \text{drt}, n)$

1    $\mathcal{X}' \leftarrow \emptyset$;

2   **foreach** $v \in \mathcal{X}$ **do**

3      $\mathcal{X}' \leftarrow \mathcal{X}' \cup N^{\text{drt}}(v) \setminus \mathcal{F}$;   // N^{drt} is N^{in} if drt = in, otherwise it is N^{out}

4   **return** $\mathcal{X}'$;

---

this amount will always lie between $|\mathcal{F}| d_{S,T} + 1$ and $|\mathcal{F}| d_{S,T} n$. The set $\mathcal{T}$ of vertices collected at the end of BFS$_\downarrow$ is called the *(target) frontier* of BFS$_\downarrow$. All vertices within $\mathcal{T}$ have the same cardinality. Also, the procedure keeps track of the *lowest* level of depth $\ell_\downarrow$ that BFS$_\downarrow$ has reached. Thus, $\ell_\downarrow$ corresponds to the *distance* between the target vertex T and the frontier $\mathcal{T}$, so that $\ell_\downarrow = |T| - |X|$ for every $X \in \mathcal{T}$. Since at the beginning of the computation, BFS$_\downarrow$ starts from the target vertex T, solve_Hy-STCON() initializes $\mathcal{T} = \{T\}$ and $\ell_\downarrow = 0$ at line 3. Figure 1 provides an illustration of double-bfs_phase().

In summary, after any round of double-bfs_phase(), we are left with two (possibly empty) frontier sets $\mathcal{S}$ and $\mathcal{T}$. In Algorithm 1, whenever $\mathcal{S} = \emptyset$ or $\mathcal{T} = \emptyset$ holds at line 6, then at least one frontier set could not proceed one level further in $\mathcal{H}_n$ while avoiding $\mathcal{F}$, and thus the procedure halts by returning NO at line 7. Similarly, whenever $\ell_\uparrow + \ell_\downarrow = d_{S,T}$ and $\mathcal{S} \cap \mathcal{T} = \emptyset$ holds at line 6, the computation halts by returning NO at line 7 — the underlying intuition being that $\mathcal{S}$ and $\mathcal{T}$ have finally reached one another's level of depth without intersecting each other, which means that $\mathcal{H}_n$ contains no directed path from S to T that avoids $\mathcal{F}^3$.

On the other hand, if both $\ell_\uparrow + \ell_\downarrow = d_{S,T}$ and $\mathcal{S} \cap \mathcal{T} \ne \emptyset$ hold at line 8, then we can prove that for every $S' \in \mathcal{S}$, there exists at least one directed path in $\mathcal{H}_n$ that goes from the source S to S' avoiding $\mathcal{F}$. Similarly, for every $T' \in \mathcal{T}$, there exists at least one directed path in $\mathcal{H}_n$ that goes from T' to target T avoiding $\mathcal{F}$. Therefore, whenever $\mathcal{S} \cap \mathcal{T} \ne \emptyset$, the algorithm is in the right position to reconstruct a directed path p in $\mathcal{H}_n$ that goes from source S to $\mathcal{S} \cap \mathcal{T}$ and from $\mathcal{S} \cap \mathcal{T}$ to target T avoiding $\mathcal{F}$ (line 9). In practice, the reconstruction can be implemented by maintaining a map throughout the computation, which associates to every vertex v (possibly visited during the BFSs) the *parent vertex*, parent(v),

---

[3] See [3], Appendix B2 for correctness.

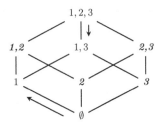

Fig. 1. A `double_bfs_phase()` on $\mathcal{H}_3$ that starts from $S = \emptyset$ and $T = \{1, 2, 3\}$. The forbidden vertices are $\mathcal{F} = \{\{2\}, \{3\}, \{1, 2\}, \{2, 3\}\}$, while the edges explored by BFS$_\uparrow$ and BFS$_\downarrow$ are $(\emptyset, \{1\})$ and $(\{1, 2, 3\}, \{1, 3\})$ (respectively).

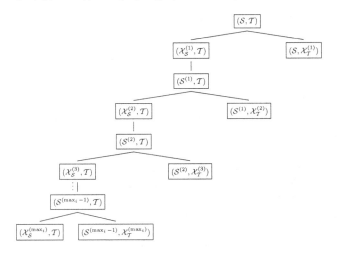

Fig. 2. The frontier sets considered during the `compression_phase()`.

which led to discover v first. As soon as p gets constructed, `solve_HY-STCON()` returns $\langle$YES, p$\rangle$ at line 10, and the computation halts.

Compression Phase. After `double-bfs_phase()` has completed, the procedure `solve_HY-STCON()` also needs to handle the case where $\mathcal{S}, \mathcal{T} \neq \emptyset$ and $\ell_\downarrow + \ell_\uparrow < d_{S,T}$. The phase that starts at that point is named `compression_phase()` (see Algorithm 3). This procedure takes as input a tuple $\langle \mathcal{S}, \mathcal{T}, \mathcal{F}, \ell_\uparrow, \ell_\downarrow, d_{S,T}, n \rangle$, where $\mathcal{S}$ and $\mathcal{T}$ are the current frontier sets. Recall that $|\mathcal{T}| > |\mathcal{F}| d_{S,T}$ holds due to line 1 of `bfs_phase()`. Also, $\mathcal{F} \subseteq \wp_n$ is the set of forbidden vertices; $\ell_\uparrow$ is the level counter of $\mathcal{S}$ and $\ell_\downarrow$ is that of $\mathcal{T}$; finally $d_{S,T}$ is the distance between the source S and the target T, and n is the size of the ground set. The output returned by `compression_phase()` is either a path p that goes from the source S to the target T avoiding $\mathcal{F}$ or a subset $\mathcal{T}' \subset \mathcal{T}$ such that the following two basic properties hold: (1) $|\mathcal{T}'| \leq |\mathcal{F}| d_{S,T}$, and (2) if p is any directed path in $\mathcal{H}_n$ that goes from S to T avoiding $\mathcal{F}$, then p goes from S to $\mathcal{T}'$.

Algorithm 3. Compression phase of $\texttt{solve\_Hy-stCon}()$.

Procedure $compression\_phase(\mathcal{S}, \mathcal{T}, \mathcal{F}, \ell_\uparrow, \ell_\downarrow, d_{\mathcal{S},\mathcal{T}}, n)$

1     $\mathcal{T}' \leftarrow \emptyset$;

2   **while** $TRUE$ **do**

3         $\mathcal{G} \leftarrow \texttt{construct\_bipartite\_graph}(\mathcal{S}, \mathcal{T}, n)$;

4         $\mathcal{M} \leftarrow \texttt{compute\_max\_matching}(\mathcal{G}, |\mathcal{F}| + 1)$;

5   **if** $|\mathcal{M}| > |\mathcal{F}|$ **then**

6            $\mathcal{M}_\mathcal{S} \leftarrow \{X \in \mathcal{S} \mid \exists Y \in \mathcal{T} \text{ s.t } (X, Y) \in \mathcal{M}\}$;

7            $\mathcal{M}_\mathcal{T} \leftarrow \{Y \in \mathcal{T} \mid \exists X \in \mathcal{S} \text{ s.t. } (X, Y) \in \mathcal{M}\}$;

8            $\{\mathsf{p}_1, \ldots, \mathsf{p}_{|\mathcal{M}|}\} \leftarrow \texttt{compute\_Lehman-Ron\_paths}(\mathcal{M}_\mathcal{S}, \mathcal{M}_\mathcal{T}, \mathcal{M}, n)$;

9            $\mathsf{p} \leftarrow \texttt{reconstruct\_path}(\mathcal{S}, \mathcal{T}, \{\mathsf{p}_i\}_{i=1}^{|\mathcal{M}|}, n)$;

10   **return** $\langle YES, \mathsf{p} \rangle$;

11        $\mathcal{X} \leftarrow \texttt{compute\_min\_vertex\_cover}(\mathcal{G}, \mathcal{M})$;

12        $\mathcal{X}_\mathcal{S} \leftarrow \mathcal{X} \cap \mathcal{S}; \mathcal{X}_\mathcal{T} \leftarrow \mathcal{X} \cap \mathcal{T}$;

13        $\mathcal{T}' \leftarrow \mathcal{T}' \cup \mathcal{X}_\mathcal{T}$;

14        $\langle \mathcal{S}, \mathcal{T}, \ell_\uparrow, \ell_\downarrow \rangle \leftarrow \texttt{double-bfs\_phase}(\mathcal{X}_\mathcal{S}, \mathcal{T}, \mathcal{F}, \ell_\uparrow, \ell_\downarrow, d_{\mathcal{S},\mathcal{T}}, n)$;

15   **if** $\mathcal{S} = \emptyset$ OR $(\ell_\downarrow + \ell_\uparrow = d_{\mathcal{S},\mathcal{T}}$ AND $\mathcal{S} \cap \mathcal{T} = \emptyset$ $)$ **then**

16   **return** $\mathcal{T}'$;

17   **if** $\ell_\uparrow + \ell_\downarrow = d_{\mathcal{S},\mathcal{T}}$ AND $\mathcal{S} \cap \mathcal{T} \neq \emptyset$ **then**

18           $\mathsf{p} \leftarrow \texttt{reconstruct\_path}(\mathcal{S}, \mathcal{T}, n)$;

19   **return** $\langle YES, \mathsf{p} \rangle$;

This frontier set $\mathcal{T}'$ is dubbed the *compression* of $\mathcal{T}$. The underlying rationale goes as follows. On one hand, because of (1), it is possible to keep the search going on by applying yet another round of $\texttt{double-bfs\_phase}()$ on input $\mathcal{S}$ and $\mathcal{T}'$ (in fact, the size of $\mathcal{T}$ has been compressed down to $|\mathcal{T}'| \leq |\mathcal{F}| \, d_{\mathcal{S},\mathcal{T}}$, thus matching the threshold condition "$|\mathcal{X}| \leq |\mathcal{F}| \, d_{\mathcal{S},\mathcal{T}}$" checked at line 1 of $\texttt{bfs\_phase}()$). On the other hand, because of (2), it is indeed sufficient to seek for a directed path in $\mathcal{H}_n$ that goes from $\mathcal{S}$ to $\mathcal{T}'$ avoiding $\mathcal{F}$, namely, the search can actually forget about $\mathcal{T} \setminus \mathcal{T}'$ because it leads to a dead end. We now describe $\texttt{compression\_phase}()$ in more details, and give a graphical summary in Fig. 2. The procedure repeatedly builds an undirected bipartite graph $\mathcal{G} = (V_\mathcal{G}, E_\mathcal{G})$, where $V_\mathcal{G} = \mathcal{S} \cup \mathcal{T}$ and every vertex $U \in \mathcal{S}$ is adjacent to a vertex $V \in \mathcal{T}$ if and only if $U \subset V$. It then uses the procedure $\texttt{compute\_max\_matching}()$ to find a matching $\mathcal{M}$ of size $|\mathcal{M}| = \min(m^*, |\mathcal{F}|+1)$, where m^* denotes the size of a maximum cardinality matching of $\mathcal{G}$. In practice, this step can be implemented in the same manner as a maximum cardinality matching procedure, e.g. as Hopcroft-Karp's algorithm [12], although with the following basic variation: if the size of the augmenting matching $\mathcal{M}$ eventually reaches the cut-off value $|\mathcal{F}| + 1$, then $\texttt{compute\_max\_matching}()$ returns $\mathcal{M}$ and halts (i.e. even if $m^* > |\mathcal{F}| + 1$). The next course of action depends on $|\mathcal{M}|$:

1. If $|\mathcal{M}| = |\mathcal{F}| + 1$, then the procedure relies on Theorem 3 to compute a family $\mathsf{p}_1, \mathsf{p}_2, \ldots, \mathsf{p}_{|\mathcal{M}|}$ of $|\mathcal{M}|$ vertex-disjoint directed paths in $\mathcal{H}_n$ that go

from S to T. In order to do that, the procedure considers the subset $\mathcal{M}_S \subseteq S$ (resp. $\mathcal{M}_T \subseteq T$) of all vertices in S (resp. in T) that are incident to some edge in $\mathcal{M}$ (lines 6 and 7). Notice that the matching $\mathcal{M}$ can be viewed as a bijection between $\mathcal{M}_S$ and $\mathcal{M}_T$. Then, the algorithm underlying Theorem 3 gets invoked on input $\langle \mathcal{M}_S, \mathcal{M}_T, \mathcal{M}, n \rangle$ (line 8). Once all the Lehman-Ron paths $\mathsf{p}_1, \mathsf{p}_2, \ldots, \mathsf{p}_{|\mathcal{M}|}$ have been found, it is then possible to reconstruct the sought directed path p in $\mathcal{H}_n$ that goes from source S to target T avoiding $\mathcal{F}$ (line 9). In fact, since $|\mathcal{M}| > |\mathcal{F}|$ by hypothesis, and since $\mathsf{p}_1, \mathsf{p}_2, \ldots, \mathsf{p}_{|\mathcal{M}|}$ are distinct and pairwise vertex-disjoint, there must exist at least one path p_i that goes from S to T avoiding $\mathcal{F}$. It is therefore sufficient to find such a path $\mathsf{p}_i = v_0 v_1 \cdots v_k$ by direct inspection. At that point, it is possible to reconstruct a path p going from S to v_0 (because $v_0 \in S$), as well as a path going from v_k to T (because $v_k \in T$). As already mentioned, in practice, the reconstruction can be implemented by maintaining a `map` that associates to every vertex v (eventually visited during the BFSs) the parent vertex that had led to discover v first. Then, $\langle \mathtt{YES}, \mathsf{p} \rangle$ is returned at line 10.

2. If $|\mathcal{M}| \leq |\mathcal{F}|$, then the `compression_phase()` aims to *compress* the size of T down to $|T'| \leq |\mathcal{F}| \, d_{S,T}$ as follows. Notice that in this case $\mathcal{M}$ is a maximum cardinality matching of $\mathcal{G}$, because $|\mathcal{M}| \leq |\mathcal{F}|$. So, the algorithm computes a minimum cardinality vertex-cover $\mathcal{X}$ of $\mathcal{G}$ at line 11, whose size is $|\mathcal{M}|$ by König's theorem [6]. The algorithm then proceeds at line 12 by considering the set $\mathcal{X}_S = \mathcal{X} \cap S$ (resp. $\mathcal{X}_T = \mathcal{X} \cap T$) of all vertices that lie both in the vertex-cover $\mathcal{X}$ and in the frontier set S (resp. T). Here, it is crucial to notice that both $|\mathcal{X}_S| \leq |\mathcal{F}|$ and $|\mathcal{X}_T| \leq |\mathcal{F}|$ hold, because $|\mathcal{X}| = |\mathcal{M}| \leq |\mathcal{F}|$. The fact that, since $\mathcal{X}$ is a vertex-cover of $\mathcal{G}$, any directed path in $\mathcal{H}_n$ that goes from S to T must go either from $\mathcal{X}_S$ to T or from $S \setminus \mathcal{X}_S$ to $\mathcal{X}_T$ plays a pivotal role. Stated otherwise, there exists no directed path in $\mathcal{H}_n$ that goes from $S \setminus \mathcal{X}_S$ to $T \setminus \mathcal{X}_T$, simply because $\mathcal{X}$ is a vertex cover of $\mathcal{G}$. At that point, the compression T' gets enriched with $\mathcal{X}_T$ at line 13.

Then, `compression_phase()` seeks a directed path in $\mathcal{H}_n$ that eventually goes from $\mathcal{X}_S$ to T. This is done at line 14 by running `double-bfs_phase()` on $\langle \mathcal{X}_S, T, \mathcal{F}, \ell_\uparrow, \ell_\downarrow, d_{S,T}, n \rangle$. Since $|\mathcal{X}_S| \leq |\mathcal{F}|$, that execution results into an update of both the frontier set S and of its level counter $\ell_\uparrow$. Let $S^{(i+1)}$ be the updated value of S and let $\ell_\uparrow^{(i+1)}$ be that of $\ell_\uparrow$. Note that, since $|T| > |\mathcal{F}| \, d_{S,T}$ holds as a pre-condition of `compression_phase()`, neither T nor $\ell_\downarrow$ are ever updated at line 14. Upon completion of this supplementary `double-bfs_phase()`, if $S^{(i+1)} = \emptyset$ or both $\ell_\uparrow^{(i+1)} + \ell_\downarrow = d_{S,T}$ *and* $S^{(i+1)} \cap T = \emptyset$ at line 15, then T' is returned at line 16 of `compression_phase()`.

Otherwise, if $\ell_\uparrow^{(i+1)} + \ell_\downarrow = d_{S,T}$ *and* $S^{(i+1)} \cap T \neq \emptyset$ at line 17, the sought directed path p in $\mathcal{H}_n$ that goes from source S to target T avoiding $\mathcal{F}$ can be reconstructed from $S^{(i+1)}$ and T at line 18, so that `compression_phase()` returns $\langle \mathtt{YES}, \mathsf{p} \rangle$ and halts soon after at line 19.

Otherwise, if $S^{(i+1)} \neq \emptyset$ and $\ell_\uparrow^{(i+1)} + \ell_\downarrow < d_{S,T}$, the next iteration will run on the novel frontier set $S^{(i+1)}$ and its updated level counter $\ell_\uparrow^{(i+1)}$. It is not difficult to prove[4] that each iteration increases $\ell_\uparrow$ by at least one unit, so that the while-loop at line 2 of compression_phase() can be iterated at most $d_{S,T}$ times overall. In particular, this fact implies that $|\mathcal{T}'| \leq |\mathcal{F}| d_{S,T}$ always holds at line 16 of compression_phase().

Figure 2 illustrates the family of all frontier sets considered throughout compression_phase(), where the following notation is assumed: $\max_i$ is the total number of iterations of the while-loop at line 2 of compression_phase(), $\mathcal{X}^{(i)}$ is the vertex-cover computed at the ith iteration of line 11, $\mathcal{X}_S^{(i)}$ and $\mathcal{X}_T^{(i)}$ are the sets computed at the ith iteration of line 12, and $\mathcal{S}^{(i)}$ is the frontier set computed at the ith iteration of line 14. The compression of $\mathcal{T}$ (possibly returned at line 16) is $\mathcal{T}' = \bigcup_{i=1}^{\max_i} \mathcal{X}_T^{(i)}$.

3.3 A Remark on Decision Versus Search

Algorithm 1 tackles the SEARCH-TASK of HY-STCON. If we merely want to answer the DECISION-TASK instead, we can simplify the algorithm by immediately returning YES if $|\mathcal{M}| > |\mathcal{F}|$ at line 5 of compression_phase(). This is because in that case, Theorem 2 guarantees the existence of a family of $|\mathcal{M}| > |\mathcal{F}|$ vertex-disjoint paths in $\mathcal{H}_n$ that go from the current source frontier $\mathcal{S}$ to the target frontier $\mathcal{T}$, which suffices to conclude that at least one of those paths avoids $\mathcal{F}$. This simplification improves the time complexity of our algorithm for solving the DECISION-TASK by a polynomial factor over that for the SEARCH-TASK.

4 Conclusion

With the intention of integrating more biologically relevant constraints into classical genome rearrangement problems, we introduced in this paper the GUIDED SORTING problem. We broadly define it as the problem of transforming two genomes into one another using as few operations as possible from a given fixed set of allowed operations while avoiding a set of nonviable genomes. We gave a polynomial time algorithm for solving this problem in the case where genomes are represented by permutations, under the assumptions that (1) permutations can only be modified by exchanging any two elements, (2) the sequence to seek must be optimal, and (3) the permutation to sort is an involution.

Many questions remain open, most notably that of the computational complexity of the GUIDED SORTING problem, whether under assumptions (1) and (2) or in a more general setting (i.e., using structures other than permutations, operations other than exchanges, or allowing sequences to be "as short as possible" instead of optimal). One could also investigate "implicit" representations for the set of forbidden intermediate permutations, e.g. all permutations that avoid a given (set of) pattern(s), or that belong to a specific conjugacy class. Aside from complexity issues, future work shall also focus on extending the approach

---

[4] See [3], Appendix B2.

we proposed to other families of instances of the GUIDED SORTING problem, and identifying other tractable (or intractable) cases or variants of it; for instance, we plan to extend our algorithmic results to the family of graphs satisfying the *shadow-matching* condition [17].

References

1. Bérard, S., Bergeron, A., Chauve, C.: Conservation of combinatorial structures in evolution scenarios. In: Lagergren, J. (ed.) RECOMB-WS 2004. LNCS (LNBI), vol. 3388, pp. 1–14. Springer, Heidelberg (2005)
2. Bergeron, A., Blanchette, M., Chateau, A., Chauve, C.: Reconstructing ancestral gene orders using conserved intervals. In: Jonassen, I., Kim, J. (eds.) WABI 2004. LNCS (LNBI), vol. 3240, pp. 14–25. Springer, Heidelberg (2004)
3. Comin, C., Labarre, A., Rizzi, R., Vialette, S.: Sorting with forbidden intermediates (extended version). CoRR abs/1602.06283 (2016)
4. Cuénot, L.: Les races pures et leurs combinaisons chez les souris. Archives de Zoologie Experimentale **3**, cxxiii–cxxxii (1905)
5. Deza, M., Huang, T.: Metrics on permutations, a survey. J. Comb., Inf. Syst. Sci. **23**, 173–185 (1998)
6. Diestel, R.: Graph Theory. Graduate Texts in Mathematics. Springer, Heidelberg (2005)
7. Fertin, G., Labarre, A., Rusu, I., Tannier, E., Vialette, S.: Combinatorics of Genome Rearrangements. The MIT Press, Cambridge (2009)
8. Figeac, M., Varré, J.-S.: Sorting by reversals with common intervals. In: Jonassen, I., Kim, J. (eds.) WABI 2004. LNCS (LNBI), vol. 3240, pp. 26–37. Springer, Heidelberg (2004)
9. Gabow, H., Maheshwari, S., Osterweil, L.: On two problems in the generation of program test paths. IEEE Trans. Softw. Eng. **2**, 227–231 (1976)
10. Gluecksohn-Waelsch, S.: Lethal genes and analysis of differentiation. Science **142**(3597), 1269–1276 (1963)
11. Goldreich, O., Goldwasser, S., Lehman, E., Ron, D., Samorodnitsky, A.: Testing monotonicity. Combinatorica **20**(3), 301–337 (2000)
12. Hopcroft, J., Karp, R.: An $n^{5/2}$ algorithm for maximum matchings in bipartite graphs. SIAM J. Comput. **2**(4), 225–231 (1973)
13. Krause, K., Smith, R., Goodwin, M.: Optimal software test planning through automated network analysis. In: Proceedings 1973 IEEE Symposium Computer Software Reliability, pp. 18–22. IEEE (1973)
14. Labarre, A.: Lower bounding edit distances between permutations. SIAM J. Discrete Math. **27**(3), 1410–1428 (2013)
15. Lakshmivarahan, S., Jwo, J.S., Dhall, S.K.: Symmetry in interconnection networks based on Cayley graphs of permutation groups: a survey. parallel comput. **19**, 361–407 (1993)
16. Lehman, E., Ron, D.: On disjoint chains of subsets. J. Comb. Theor. Ser. A **94**(2), 399–404 (2001)
17. Logan, M., Shahriari, S.: A new matching property for posets and existence of disjoint chains. J. Comb. Theor. Ser. A **108**(1), 77–87 (2004)
18. Yinnone, H.: On paths avoiding forbidden pairs of vertices in a graph. Discrete Appl. Math. **74**, 85–92 (1997)

Evaluation and Improvement of Fast Algorithms for Exact Matching on Genome Sequences

Simone Faro[(✉)]

Dipartimento di Matematica e Informatica, Università di Catania,
Viale A.Doria n.6, 95125 Catania, Italy
faro@dmi.unict.it

Abstract. With the availability of large amounts of DNA data, exact matching of nucleotide sequences has become an important application in modern computational biology and in meta-genomics. In the last decade several efficient solutions for the exact string matching problem have been developed and most of them are very fast in practical cases. However when the length of the pattern is short or the alphabet size is small (as in the case of DNA sequences) the problem becomes more difficult to be solved efficiently. In this paper we review and compare the most efficient solutions for the online exact matching problem appeared in the latest years when applied for searching on genome sequences. In addition we also propose some new variants of an efficient string matching algorithm. From our experimental results it turns out that the newly presented variants are very fast in most practical cases.

Keywords: Exact sequence analysis · String matching · Experimental algorithms · Automata based solution

1 Introduction

In molecular biology, nucleotide sequences are the fundamental information for each species and a comparison between such sequences is an interesting and basic problem. Generally biological information is stored in strings of nucleic acids (DNA, RNA) or amino acids (proteins). With the availability of large amounts of DNA data, matching of nucleotide sequences has become an important application and there is an increasing demand for fast computer methods for analysis and data retrieval [15]. There are various kinds of comparison tools which provide aligning and approximate matching (see for instance [15,19]), however most of them are based on exact matching in order to speed up the process. Moreover exact string matching is widely used in computational biology for a variety of other tasks. Thus the need for fast matching algorithms on DNA sequences.

In this article we consider the problem of finding all the (possibly overlapping) occurrences of a pattern P of length m in a text T of length n, both drawn over an alphabet Σ of size σ. We focus on the case where the text T and the pattern P are DNA sequences over a finite alphabet $\Sigma = \{a, c, g, t\}$ of constant size $\sigma = 4$. We are interested here in the problem where the pattern is given first and can

© Springer International Publishing Switzerland 2016
M. Botón-Fernández et al. (Eds.): AlCoB 2016, LNBI 9702, pp. 145–157, 2016.
DOI: 10.1007/978-3-319-38827-4_12

then be searched in various texts, thus a preprocessing phase is allowed (and in most cases suggested) on the pattern. This problem is referred in literature as the *exact online string matching problem.*

The problem of searching DNA sequences has been extensively studied in the last years and its importance in modern biology has led to produce much works. In the field of single string matching, Kalsi *et al.* [13] performed an experimental comparison of the most efficient algorithms for searching biological sequences. In addition in [8,11] Faro and Lecroq presented an extensive evaluation of (almost) all existing exact string matching algorithms (up to 2010) under various conditions, including alphabet of four characters and DNA sequences. In 2002 Navarro and Raffinot presented a comparison [18] of all matching algorithms on biological sequences, including multiple pattern matching algorithms. More recently, in 2011, Kouzinopoulos and Margaritis conducted another experimental comparison [14] taking into account the most recent solutions.

In recent years a lot of work has been made in this field and several algorithms can be considered as potential candidates to be among the fastest solutions to search genome sequences.

In this paper we present a brief survey of the most efficient solutions to the string matching problem presented in the last few years and compare them in the task of searching genome sequences. In addition we also present some efficient variants of one of the previous presented algorithms and compare them, in terms of running times, in order to evaluate their performances under various conditions. From our experimental results it turns out that some algorithms appeared in the latest years are among the fastest solutions for searching genome sequences. In addition the newly presented variants obtain the best results in almost all the practical cases.

The paper is organized as follows. In Sect. 2 we review the previous results known in literature based and describe the latest and most efficient solutions for searching genome sequences, including the BSDM algorithm. Then in Sect. 3 we present some new variants of the BSDM algorithm. In Sect. 4 we compare the newly presented solutions with the most efficient algorithms known in literature. We draw our conclusions in Sect. 5.

2   Fast Algorithms for Searching Genome Sequences

Basically a string matching algorithm uses a window to scan the text. The size of this window is equal to the length of the pattern. It first aligns the left ends of the window and the text. Then it checks if the pattern occurs in the window (this specific work is called an *attempt*) and then shifts the window to the right. It repeats the same procedure again until the right end of the window goes beyond the right end of the text.

When a similarity has been detected a naive check of the occurrence is performed. In order to detect the similarity between the pattern and the text window efficient algorithms use *bit-parallelism* or *character comparisons*. Both techniques can be improved by using condensed alphabets and hashing.

In particular the pattern P is arranged using a condensed alphabet. In such a representation groups of q adjacent characters of the pattern are condensed in a single character by using a suitable hash function $h : \Sigma^q \rightarrow \{0, \ldots, \text{MAX}\}$, for a constant value MAX. In practice, the value of q varies with m and the size of the alphabet and the value of the constant MAX varies with the memory space available.

The *bit-parallelism technique* [1] takes advantage of the intrinsic parallelism of the bit operations inside a computer word, allowing to cut down the number of operations that an algorithm performs by a factor up to ω, where ω is the number of bits in a computer word. This technique is particularly suitable for simulating non-deterministic automata for a single pattern [1] and for multiple patterns [3].

In the following sections we briefly review some of the most recent and efficient solutions for the exact string matching problem.

The Backward DAWG Matching Algorithm

One of the first application of the suffix automaton to get optimal pattern matching algorithms on the average was presented in [4]. The algorithm which makes use of the suffix automaton of the reverse pattern is called Backward-DAWG-Matching algorithm (BDM). Such algorithm moves a window of size m on the text. For each new position of the window, the automaton of the reverse of P is used to search for a factor of P from the right to the left of the window. The basic idea of the BDM algorithm is that if the backward search failed on a letter c after the reading of a word u then cu is not a factor of p and moving the beginning of the window just after c is secure. If a suffix of length m is recognized then an occurrence of the pattern was found.

The Backward Nondeterministic DAWG Matching Algorithm

The BNDM algorithm [17] simulates the suffix automaton for P^r (the reverse of P) with the bit-parallelism technique, for a given string P of length m. The bit-parallel representation uses an array B of $|\Sigma|$ bit-vectors, each of size m, where the i-th bit of $B[c]$ is set if and only if $P[i] = c$, for $c \in \Sigma$, $0 \leq i < m$. Automaton configurations are then encoded as a bit-vector D of m bits, where each bit corresponds to a state of the suffix automaton (the initial state does not need to be represented, as it is always active). In this context the i-th bit of D is set iff the corresponding state is active. D is initialized to 1^m and the first transition on character c is implemented as $D \leftarrow (D \,\&\, B[c])$. Any subsequent transition on character c can be implemented as $D \leftarrow ((D \ll 1) \,\&\, B[c])$.

The BNDM Algorithm works by shifting a window of length m over the text. Specifically, for each window alignment, it searches the pattern by scanning the current window backwards and updating the automaton configuration accordingly. Each time a suffix of P^r (i.e., a prefix of P) is found, namely when prior to the left shift the m-th bit of $D \,\&\, B[c]$ is set, the window position is recorded. A search ends when either D becomes zero (i.e., when no further prefixes of P

can be found) or the algorithm has performed m iterations (i.e., when a match has been found). The window is then shifted to the start position of the longest recognized proper prefix.

In [7] an effective variant of the BNDM algorithm was presented. Such variant, called Forward-BNDM (FBNDM), takes into account the forward character (i.e. the character which is just after the current window of the text) for computing the shift advancement. This leads to a more efficient solution especially in the case of short pattern. The FBNDM algorithm has been later improved in many ways.

The BNDM Algorithm with Extended Shift

Durian *et al.* presented in [6] another efficient algorithm for simulating the suffix automaton in the case of long patterns. The algorithm is called BNDM with eXtended Shift (BXS). The idea is to cut the pattern into $\lceil m/\omega \rceil$ consecutive substrings of length w except for the rightmost piece which may be shorter. Then the substrings are superimposed getting a superimposed pattern of length ω. In each position of the superimposed pattern a character from any piece (in corresponding position) is accepted. Then a modified version of BNDM is used for searching consecutive occurrences of the superimposed pattern using bit vectors of length ω but still shifting the pattern by up to m positions. The main modification in the automaton simulation consists in moving the rightmost bit, when set, to the first position of the bit array, thus simulating a circular automaton. Like in several other cases, the BXS algorithm works as a filter algorithm, thus an additional verification phase is needed when a candidate occurrence has been located.

The Factorized BNDM Algorithm

Cantone *et al.* presented in [2] an alternative technique, still suitable for bit-parallelism, to encode the nondeterministic suffix automaton of a given string in a more compact way. Their encoding is based on factorizations of strings in which no character occurs more than once in any factor. It turns out that the nondeterministic automaton can be encoded with k bits, where k is the size of the factorization. Though in the worst case $k = m$, on the average k is much smaller than m, making it possible to encode large automata in a single or few computer words. As a consequence, their bit-parallel variant of the BNDM, called Factorized BNDM algorithm (KBNDM) based on such approach tends to be faster in the case of sufficiently long patterns.

2.1 The Backward SNR DWAG Matching

Faro and Lecroq presented in [9] a fast and simple variant of the BDM algorithm which does not make use of bit parallelism still using a compact representation of the underlying automaton. It consist in computing the longest substring of the pattern with no repetitions, i.e. in which each character is repeated at most once, and in constructing the suffix automaton of such a substring. This leads to a simple encoding and, by convenient alphabet transformations, to quite long automata. The algorithm is named Backward-SNR-DAWG-Matching (BSDM), where SNR is the acronym of *substring with no repetitions*.

The main interesting aspect of such technique is that only an integer value between 0 and m is needed to represent the whole automaton. Since each character is repeated at most once we need only to maintain the information about the current active state, if one.

However it turns out that in many practical cases the length of the maximal SNR is not large enough if compared with the size of the pattern. This happens especially for patterns over small alphabets, as in the case of genome sequences, or for patterns with characters occurring many times, as in the case of a natural language text. In order to allow longer SNR it is convenient to use a condensed alphabet whose characters are obtained by combining groups of q characters, for a fixed value q. It turns out that the length of the maximal SNR, though quite less than m in most cases, is quite larger than the size of a computer word (which typically is 32 or 64). This leads to larger shift in a suffix automata based algorithm.

Since BSDM is a filter based algorithm, as in many other cases, a naive test is needed when a candidate occurrences of the pattern is found.

The Two-Way Shift-And Algorithm

In [5] Durian *et al.* presented the Two-Way Shift-Or algorithm (TSO) which extends the original Shift-And algorithm [1] and obtains more effective results in practical cases. Specifically it uses the same vector B as the Shift-Or algorithm but traverses the text with a fixed step of m positions. At each step i, an alignment window $T[i - m + 1...i + m - 1]$ is inspected. The positions $T[i...i + m - 1]$ correspond to the end positions of possible matches and at the same time, to the positions of the state vector D. Inspection starts at the character $T[i]$, and it proceeds with a pair of characters $T[i - j]$ and $T[i + j]$ until corresponding bits in D become 1^m or $j = m$ holds. In TSO, the testing of the state vector D is slightly faster, when the bit-vectors are seated to the highest order bits. The Two-Way Shift-And algorithm (TSA) is the dual of TSO which turns out to be faster in practical cases, especially when implemented with q enrolling characters (TSAq).

Such two-way algorithms check text in alignment windows of m consecutive text positions, thus a mismatch can be detected immediately based on the first examined text character. In the best case the performance can be $\mathcal{O}(n/m)$. On the other hand in the worst case all text characters except the last characters in each alignment window are examined twice.

The main problem associated with these solutions is that they are not able to retrive the positions of the occurrences of the pattern but only its number. A modification of such solutions which retrieve all positions of the occurrences can be obtained but with slower performance.

3   New Improvements of the BSDM Algorithm

In this section we propose some new variants of the BSDM algorithm described above, which turns out to be one of the best solutions for searching DNA sequences. Specifically we focus on reducing the number of false positives detected during the searching phase in order to reduce the number of naive tests. This can be done by using different and more effective hash function in the implementation of the condensed alphabet. In addition we use an effective technique recently introduced in [9] consisting in the use of several sliding windows while searching the pattern along the text, and which is able to speed up the whole process up to a factor of 1.3 under suitable conditions.

3.1   Improved Hash Functions for Condensed Alphabets

As we observed above, most filtering algorithms obtain better performances when used for searching sequences over large alphabets. When the size of the underlying alphabet is small it is possible to extend it by arranging the pattern P using a condensed alphabet. In such a representation groups of q adjacent characters of the pattern are condensed in a single character by using a suitable hash function $h : \Sigma^q \to \{0, \ldots, \text{MAX}\}$, for a constant value MAX. In practice, the value of q varies with m and the size of the alphabet and the value of the constant MAX varies with the memory space available.[1] Thus a pattern P of length m translates in a condensed pattern $P^{(q)}$ of length $m - q + 1$ where, for $0 \leq i \leq m - q$

$$P^{(q)}[i] = h(P[i \ldots i + q - 1]).$$

The hashing method adopted in standard implementations of condensed alphabets is based on a shift-and-addition procedure. Specifically, if $x \in \Sigma^q$, with $x = x[0 \ldots q - 1]$, then $h(x)$ can be efficiently computed by

$$h(x) = \sum_{i=0}^{q-1} ((P[i] \ \& \ M) \ll k(q - i - 1)) \tag{1}$$

where k is a constant and M is a bit-mask both dependent on q. In practice k is set to $\lfloor \omega/q \rfloor$ and M is set to $0^{\omega-k}1^k$, where ω is the size of the register used for hashing q-grams. The hash function shown in (1) has been used, for instance, in the Hashq algorithm [16] and in the Wu-Manber algorithm [20] for the exact multiple pattern matching problem.

Depending on the underlying alphabet, better hash function could be adopted in order to reduce the collisions in the hash value associated with different groups of characters. For instance the DNA alphabet is composed by the four characters $\{a, c, g, t\}$, whose ASCII codes are $\{01000001, 01000011, 01000111, 01010100\}$. Using $k = 2$ and a suitable masking leads to a perfect hashing. However for larger alphabets or when q is greater than 5 only a resemblance can be used.

---

[1] In our implementation we use a value of MAX equal to 2^{16} and use a 16-bit register for each hash value.

In our analysis we took into account six different hash functions (including the perfect hash) and evaluated them in terms of number of collisions and performances. Specifically we considered the following set of hash functions, where we set for simplicity the value of q to 4. However it is easy to extend them to greater values of the parameter q

1. Shift-Addition $(a \ll 6) + (b \ll 4) + (c \ll 2) + d$ 5 %
2. Short-Shift-Addition $(a \ll 3) + (b \ll 2) + (c \ll 1) + d$ 38 %
3. Addition $a + b + c + d$ 88 %
4. Shift-Substract $(a \ll 6) - (b \ll 4) - (c \ll 2) - d$ 5 %
5. Shift-And $(a \ll 6)$ and $(b \ll 4)$ and $(c \ll 2)$ and d 99 %
6. Shift-Or $(a \ll 6)$ or $(b \ll 4)$ or $(c \ll 2)$ or d 95 %
7. Shift-Xor $(a \ll 6)$ xor $(b \ll 4)$ xor $(c \ll 2)$ xor d 3 %
8. Perfect-Hash $(\alpha(a) \ll 6) + (\alpha(b) \ll 4) + (\alpha(c) \ll 2) + \alpha(d)$ 0 %

Where $\alpha(c) = (c$ and $6) \gg 1$, for each c in the set $\{a, c, g, t\}$.

From our analysis it turns out that the hash functions 1. and 3. obtain up to 5.5 % of collisions. When the length of the shift decreases (function 2.) the number of collisions increases to 38 % and reach the percentage of 88 % when the shift is reduced to 0. This percentage rises up to 99.2 % for hash functions 5. and 6. where the bitwise and and or are used in place of arithmetic operations. However it decreases to 3.2 % in the case of function 7. where the bitwise xor is used. Of course the number of collision is 0 in the case of function 8.

Table 1 shows the evaluation, in terms of running times, of the BSDM4 algorithm when the 8 different hash functions presented above are used. In the table running times are expressed in milliseconds and has been computed as the mean of 500 searches on a genome sequence of 5 Mb.[2] It turns out that the number of collisions generated by the hash function partially reflects the performance of the respective algorithm. However it is also affected by the number of operations needed for computing the hash value. Thus the variant using function 8.

Table 1. Experimental evaluation of the BSDM4 algorithm implemented with 8 different hash functions. Running times are expressed in milliseconds and has been computed as the mean of 500 searches on a genome sequence of 5 Mb.

m	4	8	16	32	64	128	256	512	1024	2048
FUNC.1	8.41	3.70	2.78	2.35	2.26	2.21	2.15	2.09	2.13	2.12
FUNC.2	8.82	3.95	3.03	2.63	2.49	2.46	2.38	2.38	2.35	2.36
FUNC.3	11.08	6.96	6.06	5.77	5.55	5.35	5.14	5.12	4.93	4.81
FUNC.4	8.32	3.69	2.77	2.36	2.25	2.20	2.14	2.06	2.11	2.12
FUNC.5	42.51	36.21	27.11	18.62	16.66	17.32	16.84	16.99	17.09	16.84
FUNC.6	19.58	14.27	11.24	9.09	8.01	7.45	6.73	6.44	6.10	5.98
FUNC.7	7.96	3.56	2.67	2.35	2.22	2.14	2.06	2.08	2.07	2.06
FUNC.8	10.93	4.47	3.43	2.92	2.77	2.71	2.62	2.61	2.58	2.58

---

[2] The details of experimental settings can be found in Sect. 4.

does not obtain the best results since the number of operation is doubled. Best results are obtained in all cases by the variant using the xor bitwise operation (i.e. function n.7).

3.2   A Multiple Sliding Windows Variant of the BSDM Algorithm

In this section we describe a multiple windows variant of the BSDM algorithm which improves the practical performances of the original solution. The general approach, introduced for the first time in [10], can be seen as a filtering method which consists in processing k different windows of the text at the same time, with $k \geq 2$.

Suppose P is a pattern of length m and T is a text of length n. Without loss in generality we can suppose that n can be divided by k, otherwise the rightmost ($n \bmod k$) characters of the text could be associated with the last window (as described below). Moreover we assume for simplicity that $m < n/k$ and that the value k is even.

Under the above assumptions the approach can be summarized as follows: if the algorithm searches for the pattern P in T using a text window of size m, then partition the text in $k/2$ partially overlapping substrings, $T_0, T_1, \ldots, T_{k/2-1}$, where T_i is the substring $T[2i\lceil n/k \rceil \mathinner{..} 2(i+1)n/k+m-2]$, for $i = 0, \ldots, (k-1)/2$, and $T_{k/2-1}$ (the last window) is set to $T[n - (2n/k) \mathinner{..} n - 1]$.

Then process simultaneously the k different text windows, $w_0, w_1, \ldots, w_{k-1}$, where we set $w_{2i} = T[s_{2i} - m + 1 \mathinner{..} s_{2i}]$ (and call them *left windows*) and $w_{2i+1} = T[s_{2i+1} \mathinner{..} s_{2i+1} + m - 1]$ (and call them *right windows*), for $i = 0, \ldots, (k-2)/2$.

The couple of windows (w_{2i}, w_{2i+1}), for $i = 0, \ldots, (k-2)/2$, is used to process the substring of the text T_i. Specifically the window w_{2i} starts from position $s_{2i} = (2n/k)i + m - 1$ of T and slides from left to right, while window w_{2i+1} starts from position $s_{2i+1} = (2n/k)(i+1) - 1$ of T and slides from right to left (the window w_{k-1} starts from position $s_{k-1} = n - m$). For each couple of windows (w_{2i}, w_{2i+1}) the sliding process ends when the window w_{2i} slides over the window w_{2i+1}, i.e. when $s_{2i} > s_{2i+1} + m - 1$. It is easy to prove that no candidate occurrence is left by the algorithm due to the $m - 1$ overlapping characters between adjacent substrings t_i and t_{i+1}, for $i = 0, \ldots, k - 2$.

Fig. 1 presents a scheme of the search iteration of the multiple sliding windows matcher for $k = 1, 2$ and 4. It has been proved in [10] that this approach can be applied to all string matching algorithms, including the BSDM algorithm. Moreover it can be noticed that the worst case time complexity of the original algorithm does not degrade with the application of the multiple sliding windows approach.

On the other hand it turns out, when the alphabet is small as in the case of DNA sequences, that the performances of the original algorithm degrade by applying the new method, since the probability to find mixed candidate positions increases substantially.

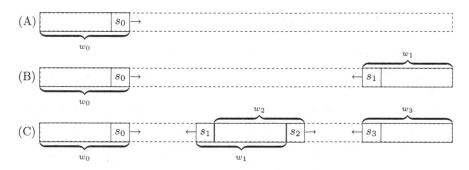

Fig. 1. A general scheme for the multiple sliding windows approach with (A) a single window, (B) two windows and (C) four windows (w_1 and w_2 are overlapping).

4 Experimental Results

In this section we briefly present experimental evaluations in order to understand the performances of the newly presented algorithm and to compare it against the best string matching algorithms for searching genome sequences. In particular we tested the following algorithms:

- the Backward-Nondeterministic-DAWG-Matching algorithm [17] (BNDMq) implemented using q-grams and a value of $q = 4$;
- the Extended Backward-Oracle-Matching algorithm [7] (EBOM);
- the Hashing algorithm [16] (HASHq) implemented using q-grams and a value of $q \in \{3, 4, 5\}$;
- the Simplified version of the BNDM algorithm [17] (SBNDMq) implemented using q-grams and a value of $q = 4$;
- the Forward Simplified version of the BNDM algorithm [7] (FSBNDMq) implemented using q-grams and a value of $q = 4$;
- the Multiple Windows version of the Forward Simplified BNDM algorithm [10] (FSBNDM-W4) implemented using 4 sliding windows;
- the Factorized BNDM algorithm [2] (KBNDM);
- the BNDM algorithm with Extended Shift [6] (BXSq) implemented using q-grams and a value of $q = 4$;
- The Backward-SNR-DAWG-Matching algorithm [9] using condensed alphabets with groups of q characters, with $q \in \{1, 2, 4, 6, 7\}$ (BSDMq);
- The new BSDM algorithm using condensed alphabets and a *shift-xor* hash function, with $q \in \{1, 2, 4, 6, 7\}$ (BSDMqx);
- The Multiple Windows version of the new BSDM algorithm using condensed alphabets and a *shift-and-xor* hash function (BSDMqx-w2 and BSDMqx-w4) implemented using 2 and 4 sliding windows, respectively;

For the sake of completeness we evaluate also the following two string matching algorithms for *counting* occurrences. They do not report the positions but only the total number of all occurrences.

- EPSM: the Exact Packed String Matching algorithm [12];
- TSOq: the Two-Way variant of [5] the Shift-Or algorithm [1] implemented with a loop unrolling of q characters, with $q = 5$;

All algorithms have been implemented in the C programming language and have been tested using the SMART tool[3]. The experiments were executed locally on an MacBook Pro with 4 Cores, a 2 GHz Intel Core i7 processor, 16 GB RAM 1600 MHz DDR3, 256 KB of L2 Cache and 6 MB of Cache L3. Algorithms have been compared in terms of running times, including any preprocessing time.

For the evaluation we use the genome sequence provided by the SMART research tool. Specifically it is a sequence of $4,638,690$ base pairs of *Escherichia coli*, maintained by the Large Canterbury Corpus.[4] In all cases the patterns were

Table 2. Experimental results on a genome sequence. Best results have been bold faced. Running times are expressed in milliseconds. For the algorithms using variable q-grams we report in brackets the value of q which obtains the best running times. The EPSM and TSOq algorithms (indicated by a * symbol) are *counting* algorithm, i.e. it is able only to count occurrences.

m	4	8	16	32	64	128	256	512	1024	2048
BNDMq	$11.14^{(4)}$	$4.12^{(4)}$	$3.02^{(4)}$	$2.41^{(4)}$	$2.41^{(4)}$	$2.39^{(4)}$	$2.23^{(4)}$	$2.40^{(4)}$	$2.32^{(4)}$	$2.33^{(4)}$
EBOM	**7.74**	7.00	5.53	4.01	3.10	2.62	2.38	2.21	2.36	2.65
HASHq	$18.31^{(3)}$	$7.67^{(3)}$	$4.69^{(5)}$	$3.32^{(5)}$	$2.85^{(5)}$	$2.35^{(5)}$	$2.57^{(5)}$	$2.45^{(5)}$	$2.34^{(5)}$	$2.30^{(5)}$
SBNDMq	$10.32^{(4)}$	$4.00^{(4)}$	$2.96^{(4)}$	$2.37^{(4)}$	$2.39^{(4)}$	$2.31^{(4)}$	$2.29^{(4)}$	$2.34^{(4)}$	$2.28^{(4)}$	$2.37^{(4)}$
BSDMq	$8.43^{(4)}$	$3.02^{(6)}$	$2.41^{(6)}$	$2.42^{(7)}$	$2.13^{(7)}$	$1.98^{(7)}$	$2.01^{(7)}$	$2.00^{(7)}$	$2.00^{(7)}$	$2.00^{(7)}$
BXSq	$15.34^{(4)}$	$4.50^{(4)}$	$3.11^{(4)}$	$2.39^{(4)}$	$2.42^{(4)}$	$2.37^{(4)}$	$2.40^{(4)}$	$2.40^{(4)}$	$2.40^{(4)}$	$2.41^{(4)}$
FS-w4	16.54	5.20	5.05	3.76	4.09	3.61	3.88	3.52	3.34	3.09
FSBNDM-w4	17.10	8.09	4.30	2.99	3.00	2.89	2.94	2.94	2.97	2.95
KBNDM	10.80	8.00	5.88	3.98	3.03	2.91	2.94	2.92	2.97	2.97
TSOq (*)	$\mathbf{5.32}^{(5)}$	$3.68^{(5)}$	$2.88^{(5)}$	$2.28^{(5)}$	$1.96^{(5)}$	-	-	-	-	-
EPSM (*)	5.87	3.72	2.50	**1.93**	**1.75**	**1.72**	**1.66**	**1.62**	**1.65**	**165**
BSDM2x	8.32	7.53	6.80	6.12	5.66	5.24	4.94	4.70	4.46	4.28
BSDM4x	8.09	**3.01**	**2.40**	2.21	2.07	2.03	1.96	1.96	1.94	1.94
BSDM6x	-	4.79	3.02	2.43	2.15	2.04	2.01	2.00	1.98	2.00
BSDM7x	-	6.85	3.10	2.35	2.07	1.96	1.92	1.92	1.92	1.97
BSDM2x-w2	8.94	7.25	6.57	5.82	5.39	5.03	4.70	4.48	4.24	4.09
BSDM2x-w4	9.51	7.27	6.66	5.91	5.52	5.08	4.66	4.42	4.19	4.05
BSDM4x-w2	10.22	3.22	2.54	**2.16**	2.01	1.96	1.91	1.89	1.86	1.90
BSDM4x-w4	12.72	3.59	2.66	2.25	2.11	2.02	1.96	1.94	1.92	1.95
BSDM6x-w2	-	5.94	3.08	2.40	2.14	2.01	1.99	1.95	1.91	1.97
BSDM6x-w4	-	9.32	3.30	2.36	2.07	1.88	1.85	**1.83**	**1.81**	**1.84**
BSDM7x-w2	-	8.41	3.18	2.29	**1.99**	**1.83**	**1.82**	**1.83**	**1.81**	**1.84**
BSDM7x-w4	-	7.47	3.01	2.30	2.02	1.86	1.85	**1.83**	1.82	1.88

[3] SMART, a String Matching Algorithms Research Tool, by Simone Faro and Thierry Lecroq, http://www.dmi.unict.it/~faro/smart.
[4] http://www.data-compression.info/Corpora/CanterburyCorpus/.

randomly extracted from the text and the value m was made ranging from 4 to 2048. For each case we reported the mean over the running times of 500 runs.

From experimental results reported in Table 2 it turns out that the BSDMq algorithm obtains the best results in almost all the cases. In particular the best running times are obtained with a value of q equal to 7 (for long patterns) and 4 (for short patterns).

Comparing the new presented algorithms against the previous known solutions we can observe that the new BSDMq-w algorithms are the fastest in most cases, especially for long patterns. We can observe moreover that when the length of the pattern gets longer, better results are obtained for greater values of q. The same observation could be done for the number of windows used during the searching.

Specifically the BSDM algorithm implemented with 7-grams and two sliding windows obtains the best results when the pattern is longer than 32 characters. In this cases the BSDMq-w is up to 10 % faster than BSDMq and up to 13 % faster than all previous known algorithms.

When the size of the pattern is between 8 and 32 the new BSDMqx algorithms obtain the best results, using 4-grams. In this cases the BSDM4x algorithm is up to 1, 2 times faster than the best among the previous solutions (the SBNDMq algorithm).

For small patterns ($m = 4$) the best running time is obtained by the EBOM algorithm, even if we have to report the good performance of the TSO5 algorithm, although it's not able to report the positions of the found occurrences. However, it is interesting to observe that when $m = 4$ the BSDM4x algorithm obtains a result which is very close to the best running time.

5 Conclusions and Future Works

In this paper we reviewed the most recent and efficient solutions for searching exact matching on genome sequences. We also compared such solutions in terms of running times in order to identify the best solutions for such problem. In addition we also propose some efficient variants of the BSDM algorithm which turn out to be competitive with the previous solutions and obtain the best running times in most practical cases.

From experimental results it turns out that the new presented variants obtain the best results in most practical cases when tested join real genome sequences.

It will be interesting to investigate if some of the efficient solutions described above could be generalized also in the case of multiple pattern matching on genome sequences.

Acknowledgments. This work has been supported by G.N.C.S., Istituto Nazionale di Alta Matematica "Francesco Severi".

References

1. Baeza-Yates, R., Gonnet, G.H.: A new approach to text searching. Commun. ACM **35**(10), 74–82 (1992)
2. Cantone, D., Faro, S., Giaquinta, E.: A compact representation of nondeterministic (suffix) automata for the bit-parallel approach. In: Amir, A., Parida, L. (eds.) CPM 2010. LNCS, vol. 6129, pp. 288–298. Springer, Heidelberg (2010)
3. Cantone, D., Faro, S., Giaquinta, E.: A compact representation of nondeterministic (suffix) automata for the bit-parallel approach. Inform. Comput. **213**, 3–12 (2012)
4. Crochemore, M., Rytter, W.: Text Algorithms. Oxford University Press, New York (1994)
5. Durian, B., Chhabra, T., Ghuman, S., Hirvola, T., Peltola, H., Tarhio, J.: Improved two-way bit-parallel search. In: Stringology Conference 2014, pp. 71–83 (2014)
6. Durian, B., Peltola, H., Salmela, L., Tarhio, J.: Bit-parallel search algorithms for long patterns. In: Festa, P. (ed.) SEA 2010. LNCS, vol. 6049, pp. 129–140. Springer, Heidelberg (2010)
7. Faro, S., Lecroq, T.: Efficient variants of the backward-oracle-matching algorithm. In: Prague Stringology Conference, pp. 146–160. Czech Technical University in Prague, Czech Republic (2008)
8. Faro, S., Lecroq, T.: The exact string matching problem: a comprehensive experimental evaluation. CoRR abs/1012.2547 (2010)
9. Faro, S., Lecroq, T.: A fast suffix automata based algorithm for exact online string matching. In: Moreira, N., Reis, R. (eds.) CIAA 2012. LNCS, vol. 7381, pp. 149–158. Springer, Heidelberg (2012)
10. Faro, S., Lecroq, T.: A multiple sliding windows approach to speed up string matching algorithms. In: Klasing, R. (ed.) SEA 2012. LNCS, vol. 7276, pp. 172–183. Springer, Heidelberg (2012)
11. Faro, S., Lecroq, T.: The exact online string matching problem: a review of the most recent results. ACM Comput. Surv. **45**(2), 13 (2013)
12. Faro, S., Kulekci, M.O.: Fast and flexible packed string matching. J. Discrete Algorithms **28**, 61–72 (2014)
13. Kalsi, P., Peltola, H., Tarhio, J.: Comparison of exact string matching algorithms for biological sequences. In: Elloumi, M., Küng, J., Linial, M., Murphy, R.F., Schneider, K., Toma, C. (eds.) Bioinformatics Research and Development, pp. 417–426. Springer, Heidelberg (2008)
14. Kouzinopoulos, C.S., Michailidis, P.D., Margaritis, K.G.: Experimental results on multiple pattern matching algorithms for biological sequences. In: Bioinformatics, pp. 274–277 (2011)
15. Langmead, B., Trapnell, C., Pop, M., Salzberg, S.: Ultrafast and memory-efficient alignment of short dna sequences to the human genome. Genome Biol. **10**(3), 5–10 (2009)
16. Lecroq, T.: Fast exact string matching algorithms. Inf. Process. Lett. **102**(6), 229–235 (2007)
17. Navarro, G., Raffinot, M.: A bit-parallel approach to suffix automata: fast extended string matching. In: Farach-Colton, M. (ed.) CPM 1998. LNCS, vol. 1448, pp. 14–33. Springer, Heidelberg (1998)
18. Navarro, G., Raffinot, M.: Flexible Pattern Matching in Strings: Practical Online Search Algorithms for Texts and Biological Sequences. Cambridge University Press, New York (2002)

19. Rivals, E., Salmela, L., Kiiskinen, P., Kalsi, P., Tarhio, J.: MPSCAN: fast localisation of multiple reads in genomes. In: Salzberg, S.L., Warnow, T. (eds.) WABI 2009. LNCS, vol. 5724, pp. 246–260. Springer, Heidelberg (2009)
20. Wu, S., Manber, U.: A fast algorithm for multi-pattern searching. Report TR-94-17, Depart. of Computer Science, University of Arizona, Tucson, AZ (1994)

Identification of Variant Compositions in Related Strains Without Reference

Mikko Rautiainen, Leena Salmela, and Veli Mäkinen[(✉)]

Department of Computer Science, Helsinki Institute for Information Technology,
University of Helsinki, Helsinki, Finland
m_rautiainen@hotmail.com, {lmsalmel,vmakinen}@cs.helsinki.fi

Abstract. Current DNA sequencing technologies do not read an entire chromosome from end to end but instead produce sets of short *reads*, i.e. fragments of the genome. *Haplotype assembly* is the problem of assigning each read to the correct chromosome in the set of chromosomes in a homologous group, with the aid of the reference sequence. In this paper, we extend an existing exact algorithm for haplotype assembly of diploid species (Patterson et al., 2014) to the *reference-free*, polyploid case. A reference-free method does not exploit a reference genomic sequence of a species and thus we cannot exploit a known linear order for the reads and resulting variant positions. Therefore we obtain an unordered *variant composition* as a result. This setting can be also applied to the study of relative abundances of related bacterial strains.

Keywords: Variant composition · Reference free sequence analysis

1 Introduction

The *genome* of an organism is a sequence or sequences of the nucleotides, A, T, C, and G. Eukaryotic genomes are composed of *chromosomes*, each of which is a sequence of nucleotides. The chromosomes are further arranged into groups of *homologous* chromosomes, where two or more chromosomes are almost exact copies of each other. For example, humans' genomes are arranged into pairs of homologous chromosomes, one of which is inherited from the mother and the other from the father. Species with pairs of homologous chromosomes are called *diploid*. Species with groups of more than two homologous chromosomes, for example some potato species which have groups of 3 or 4 chromosomes [7], are called *polyploid*. The sequence of a chromosome is called a *haplotype sequence*, and a chromosome in a homologous group is sometimes called a *haplotype*.

Homologous chromosomes are typically very similar to each other with minor variation. A common variation is the *single nucleotide polymorphism* (SNP), where the haplotype sequences vary by exactly one nucleotide at some location, either as a substitution, having a different nucleotide in the chromosomes, or indels (insertion-deletion), where a nucleotide has been inserted or deleted from one of the chromosomes. SNPs can be either *heterozygous*, where all of the

M. Botón-Fernández et al. (Eds.): AlCoB 2016, LNBI 9702, pp. 158–170, 2016.
DOI: 10.1007/978-3-319-38827-4_13

haplotype 1   ACT**A**ACGC**T**GAAGAC**T**AGT
haplotype 2   ACT**C**ACGC**T**GAAGAC - AGT   ACT^A_CACGCTGAAGAC$^T_-$AGT
reference   ACT**A**ACGC**A**GAAGAC**G**AGT

(a) Haplotypes and reference sequence   (b) Consensus sequence

Fig. 1. The haplotypes of a genome and the corresponding consensus sequence

haplotypes have a different variant (nucleotide), or *homozygous*, where some haplotypes have the same variant. Figure 1a shows an example of SNPs. Two chromosomes from an organism are compared to a *reference genome*. The sequences have three SNPs. In the beginning there is a heterozygous SNP with a substitution, in the middle there is a homozygous SNP with substitutions, and in the end there is a heterozygous SNP with an indel and a substitution.

Current sequencing technologies produce sets of *reads*, also called *fragments*, which are short sequences of DNA from anywhere in the organism's genome. Due to the difficulty of assembling a genome from sequenced reads, most species lack a reference genome. A reference genome may also be incomplete or otherwise unusable. Furthermore, polyploid genomes are more difficult to sequence than diploid sequences. Another similar case is a population of closely related bacteria or virii. A reference genome may not be available given the sheer amount of bacterial species. For this reason, *reference-free* algorithms which do not require a reference genome are useful for poorly understood genomes.

Since the chromosomes or strains are very similar, they can be represented as one sequence and a list of positions where SNPs are found. Figure 1b shows an example with two sequences having a common consensus with two SNPs. However, the consensus does not tell whether the A and the T belong in the same sequence or not. SNPs have different effects on an organism's phenotype depending on which other variants are found in the same chromosome [19,21], thus having just a list of SNPs misses *phase* information, which describes which variants occur in the same chromosomes or strains. Identification of *variant compositions* aims to correct this. A variant composition is an assignment of the variants to the chromosomes or strains. The process of assigning the variants of a SNP to the chromosomes is called *phasing*. In this paper we generalize an existing exact phasing algorithm for diploid genomes [17] to the reference-free, polyploid case. The phasing algorithm's time and space complexities are both exponential in the maximum *coverage* of the input. The coverage of a position is the number of reads that cover the position. See Fig. 2 for an example.

2   Previous Work and Our Contribution in Context

Identification of variant compositions is closely related to the *haplotype assembly problem* which seeks to solve the sequences of an organism's haplotype. A variant composition is a list of variants for each haplotype at the SNPs, which can be used to calculate the haplotype sequences given a reference genome. Haplotype

Fig. 2. The long, solid line is the genome, which is unknown, and the short solid lines represent the reads and their location in the genome. The position at the first dashed line is covered by three reads and thus has a coverage of three. Similarly, the second dashed line has a coverage of five

assembly problem has multiple similar formulations [14]. These formulations are based on building a *haplotyping matrix*, where the reads are rows and SNPs are columns, and partitioning the rows into chromosomes such that all reads in a partition agree on the variant at each SNP. A haplotyping matrix does not always have such a partition, and then the problem is to modify the matrix the least to make such a partition possible. Lippert et al. [14] suggest three formulations, of which we follow the *minimum error correction* that aims to flip the least amount of cells in the matrix. Haplotype assembly is NP-hard both in general [14] and also in the case when there are no gaps between reads [6].

Identification of variant compositions is also related to the *quasispecies spectrum reconstruction problem* [2], which aims at reconstructing the sequences of the strains in a population of bacteria or virii and their frequencies. Variant compositions can be used to reconstruct the sequences, once reference sequences are given.

Haplotype assembly for diploid organisms, organisms only having two chromosomes, has many algorithms available, both exact [5,9,10,17] and probabilistic [3,12]. Solving the haplotype assembly for polyploid organisms, organisms having more than two chromosomes, is a more recent field of research. Both exact [8,16] and probabilistic algorithms [1,4,20] are available.

Deng et al. [9] published in 2013 an exact algorithm for the diploid haplotype assembly problem with the minimum error correction formulation. The algorithm has a time complexity of $O(|S|2^C C)$ where $|S|$ is the number of SNPs and C is the maximum coverage. Patterson et al. [17] extended the algorithm to the weighted minimum error correction formulation, and improved the time complexity to $O(|S|2^{C-1})$. In this paper we further extend the algorithm to the polyploid case with a time complexity of $O(|S|C\frac{k^C}{k!})$, where k is the number of chromosomes. The algorithm uses only the reads and requires no reference genome.

For our implementation, we use DiscoSNP [22] to detect SNPs in the reads in a reference-free manner. DiscoSNP first builds a de Bruijn graph of the reads and then uses it to find the SNPs.

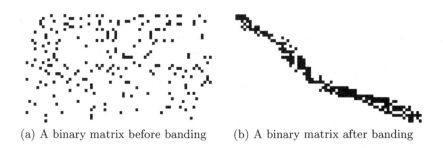

(a) A binary matrix before banding   (b) A binary matrix after banding

Fig. 3. Matrix banding

3 Extending Haplotyping to Multiple Strains Without Reference

Our method has three stages. As preprocessing, a *haplotyping matrix* is built from the reads. The second stage manipulates the haplotyping matrix to lower coverage. The third stage, the *phasing* stage, assigns the reads into strains. Due to the lack of space, we describe the coverage reduction and phasing stages that form the algorithmic core of our contribution. More details are available in [18].

3.1 Coverage Reduction

After the preprocessing stages, we assume to have a haplotyping matrix F that contains the SNPs as columns, reads as rows, and the value of a cell is either the read's nucleotide at the SNP or a marker "$-$" indicating that the read does not cover that SNP. In the first case, the read is said to *support* a certain variant, that is the nucleotide, at the SNP. The haplotyping matrix is then said to have a variant at that cell. Additionally we have a weight matrix W where the entry $W(i, x)$ describes the certainty of the haplotyping matrix entry $F(i, x)$.

Because the phasing algorithm is exponential in coverage, reducing the coverage is necessary. Coverage is either *essential*, when a read supports some variant for a SNP, or *accidental* when a read does not directly support a SNP, but supports SNPs both before and after it.

Since a reference genome is not used, the reads cannot be ordered by aligning them to a reference genome. Instead we use matrix banding [11] to reduce accidental coverage. The haplotyping matrix is treated as a binary matrix, where cells with a variant are 1, and cells without a variant are 0. The rows and columns are then permutated to bring the ones together in the matrix. The same permutation is also applied to the weight matrix. Figure 3 shows an example of a matrix before and after banding. Cells with a variant are colored black, and cells without are white. The ideal output would be a solid diagonal band.

We use methods presented in [11] for matrix banding. First we find an approximate *consecutive-ones property* (C1P) on the haplotyping matrix. The C1P is a permutation of the columns such that in any row, the ones are in consecutive positions. Then *barycentric sorting* is used. For each row, a centerpoint is calculated as the average position of the ones and the rows are sorted based on the centerpoints. The same operation is then performed on the columns. This is repeated for a certain number of iterations and the best iteration's matrix is selected. The last step in matrix banding uses *simulated annealing*.

After banding, methods similar to the ones in [15] are used to still lower the essential coverage.

3.2   Phasing

The algorithm by Patterson et al. [17] works with two strains. Here we extend the algorithm for an arbitrary number of strains. The output of the phasing algorithm is an assignment of the reads into the strains.

A *partition* is an assignment of the reads, or a subset of reads, to the strains. Partitions are stored as a *partition vector*, which describes for each read either the strain where it is assigned, or a marker that this read is not assigned to any strain. For example, $\{1, -, 2, 1, 3, -\}$ is a partition of four reads into three strains, where reads 1 and 4 are in the same strain and reads 2 and 6 are not assigned to any strain. The example partition is over reads 1, 3, 4, and 5.

The overlap of two partitions is the indices where they both have a value. For example, the overlap of $\{1, 2, 2, -, -\}$ and $\{-, 1, 2, 3, 3\}$ is indices 2 and 3.

The *unpermutated form* of a partition P is marked as $u(P)$. In the unpermutated form the strains are re-labeled according to the order in which they appear in the partition vector. For example, the partition vector $\{3, 1, 3, 2, 2, 1, 4\}$ has the unpermutated form $\{1, 2, 1, 3, 3, 2, 4\}$.

This operation of re-labeling the strains is called a *renumbering* which is a bijection on strain numbers. A *renumbering vector* describes how the strains are re-labeled. Given a partition P and a renumbering vector R, the renumbered partition P' is given by $P'_i = R_{P_i}$. The renumbering vector for the example above is $\{2, 3, 1, 4\}$. For example, strain number 2 in the original partition is replaced with 3 in the renumbered partition. Changing a partition to its unpermutated form is one example of a renumbering but others are also used in the algorithm.

Two partitions are equivalent if and only if their unpermutated forms are equal, and then both partitions assign the reads into same sets. Only partitions in the unpermutated form need to be considered in the algorithm since all other partitions are permutations of some unpermutated form.

The set of *active reads* for a SNP is the reads that cover it, either essentially or accidentally. The active reads for SNP i are denoted as $\alpha(i)$.

The set of all partitions over a set of reads r is marked as $Par(r)$. Note that this set does not actually have all partitions, only all unpermutated forms. This cuts the number of partitions from $k^{|r|}$ to $\frac{k^{|r|}}{k!}$.

A partition P_1 *extends* partition P_2 if and only if their unpermutated forms over the overlapping area are equal. The notation $Ext(P_1, P_2)$ is used to mark this. The set of partitions that extend a partition is said to be its *extensions*.

A partition is *conflict-free* if all reads in a strain assign the same variant at each SNP. For example, the haplotyping matrix in Fig. 4 is conflict-free for the partition vector $\{1, 2, 1, 2\}$, but not for the partition vector $\{1, 2, 2, 2\}$. In the conflicting example, the cell at read 3 and SNP 2 is in conflict since it is different from the consensus variant of strain 2 for SNP 2.

	1	2
1	C	T
2	C	C
3	C	T
4	C	C

Fig. 4. A haplotyping matrix

In practice, the haplotyping matrix rarely has a conflict-free partition. Instead, the algorithm finds the partition closest to a conflict-free partition, with the distance being the total weight of the conflicting cells in the haplotyping matrix. For a partition P, haplotyping matrix F, weight matrix W, strain s, variant $v \in \{A, T, C, G\}$, and a SNP number i, we define a cost function

$$\delta(P, s, v, i) = \sum_{x \mid P_i = s \wedge F(x,i) \neq v} W(x, i).$$

The function describes the cost for assigning the strain s to have variant v at SNP i. Using this, we define the *partition cost function*

$$\Delta(P, i) = \sum_{s \in [1,k]} \min_{v \in \{A,T,C,G\}} \delta(P, s, v, i)$$

as the cost of partition P at SNP i. This cost function is the minimum cost to make the partition conflict-free. These functions are generalizations of the cost functions described in [17], and are equal to them in the case of two strains.

The algorithm is a dynamic programming algorithm that goes through the SNPs one by one. The table C contains the best scores for a partition at any SNP; element $C(i, x)$ is the best score at SNP i for partition x. At the first SNP, the table is initialized with

$$C(1, x) = \Delta(x, 1), \forall x \in Par(\alpha(1)).$$

At every SNP other than the first, the cost of a partition at the current SNP is

$$C(i, x) = \Delta(x, i) + \min_{y \in Par(\alpha(i-1)), Ext(x,y)} C(i - 1, y).$$

which is the sum of the cost for that partition at the current SNP and the cost of the best-scoring partition of the previous SNP that is extended by the current partition. To get the final result, the table C can be backtraced and the partitions at each SNP must be merged.

When two partitions are merged, one of them must usually be renumbered. For example, the partition $\{-, -, 1, 2, 1, 1, 3\}$ extends the partition $\{1, 2, 3, 4, 3, -, -\}$, since the unpermutated form of the overlapping part is $\{-, -, 1, 2, 1, -, -\}$ for both partitions. The algorithm renumbers the rightmost partition to correspond to the leftmost partition's strain numbers. The renumbering vector is constructed by first assigning the overlapping part. Having a left partition P, a right partition Q and a set of indices for the overlapping part I, the renumbering vector R is given by the equation

$$R_{P_x} = Q_x, x \in I$$

This equation only assigns the renumbering vector for the strains which appear in the overlapping part and it works only when the two partitions actually extend each other, otherwise it would produce more than one value for some indices. Some indices may be left unassigned by this equation. The remaining indices must be arbitrarily chosen to make the renumbering vector a bijection. The implementation simply assigns the remaining values in order. In the example above, the renumbering vector would be $\{3, 4, 1, 2\}$. The values of the first two indices are determined by the overlapping part, and the last two arbitrarily. The final merged partition is then $\{1, 2, 3, 4, 3, 3, 1\}$. After the algorithm has passed through all SNPs, the solution is the merged partition with the lowest score.

When some values of the renumbering vector are chosen arbitrarily, the merged partition's early reads and late reads are assigned essentially randomly. In the above example another consistent merging is $\{1, 2, 3, 4, 3, 3, 2\}$. Therefore some strains may be swapped in the middle of the partition. The experiments section measures how often this happens in practice.

Calculating the partition cost function δ directly from its equation would take $O(c)$ time, where c is the coverage at the current SNP, and Δ would take $O(kc)$ time. However, two optimizations make it possible to do it faster. The first optimization is to iterate through the partition only once, and calculate δ for all strains simultaneously. This is done by keeping a two-dimensional array x with size $k * 4$ that keeps track of the cost of assigning a strain to a nucleotide. At each read in the partition, the cost of the strain that the read is assigned to is increased at the nucleotide the read has at that position. For example, if read 5 is assigned to strain 2, and read 5 has the nucleotide A at the current SNP, then the value at $x_{2,A}$ is increased by the weight of the cell when the iteration handles read 5. The final cost is then calculated with

$$\Delta(P, i) = \sum_{s \in [1..k]} \max(x_{s,A}; x_{s,T}; x_{s,C}; x_{s,G}) - (x_{s,A} + x_{s,T} + x_{s,C} + x_{s,G})$$

This improves the cost for calculating Δ to $O(c + k)$.

The second optimization uses a *Gray code* to order the partitions. A Gray code is an ordering of vectors, in this case the partition vectors, where two consecutive vectors differ by exactly one element. Then, calculating the next x from the previous x can be done in constant time. At each SNP, the first x must be calculated as usual. Then, using a notation $x_{i,s,a}$ to mark x for the ith

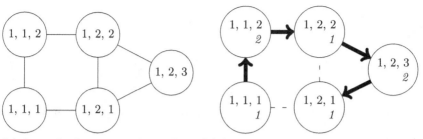

(a) A graph of partitions for 3 reads   (b) A hamiltonian path in the partition graph over 3 reads with nodes marked

Fig. 5. Example of partitions for three reads

partition, $x_{i+1,s,a}$ is equal to $x_{i,s,a}$ except for the element that changed between the two partitions. For example, if the read 5 with nucleotide A was assigned to strain 3 at the previous partition, and to strain 4 at the current partition, then the next x is calculated with

$$\begin{cases} x_{i+1,3,A} = x_{i,3,A} - W(5,i) \\ x_{i+1,4,A} = x_{i,4,A} + W(5,i) \\ x_{i+1,s,a} = x_{i,s,a} \qquad\qquad \text{otherwise} \end{cases}$$

This removes the need to iterate through each partition and Δ can be calculated in amortized constant time. The Gray code optimization was originally described by Patterson et al. [17]. However, extending it to multiple strains requires a special Gray code that orders only the unpermutated forms of the partitions.

Consider a graph of the partitions, where nodes are partitions, and edges connect partitions which differ by exactly one element. A Gray code over the partitions is possible if the graph has a Hamiltonian path, or a path visiting each node exactly once. For a partition over one read, this is trivially true. Then, a path for the graph with $c + 1$ reads can be constructed from a path in the graph with c reads. Each node in the graph with c reads is divided into the partitions where the first c elements are equal, and the final element varies. Figures 5a and 6 show an example of extending a graph of 3 reads to a graph of 4 reads. The node $\{1, 2, 1\}$ is divided into the nodes $\{1, 2, 1, 1\}$, $\{1, 2, 1, 2\}$ and $\{1, 2, 1, 3\}$.

There are two key features of the graph that make a Hamiltonian path possible. First, a node's child nodes form a clique, so they can be visited in any order. Second, if two nodes were connected in the previous graph, their child cliques will have at least two connections between the cliques: the nodes that end with 1 are connected, and the nodes that end with 2 are connected. These nodes exist in all cliques. Therefore, to build a Hamiltonian path for the $c + 1$ graph from the c graph, mark every other node in the path with a 1, and every other with a 2. Then, for each node that is marked with a 1, the sub-path starts at the child clique's node that ends in 1, visits all nodes that end in 3 or higher, and ends at the child node that ends with 2. Correspondingly, for nodes marked

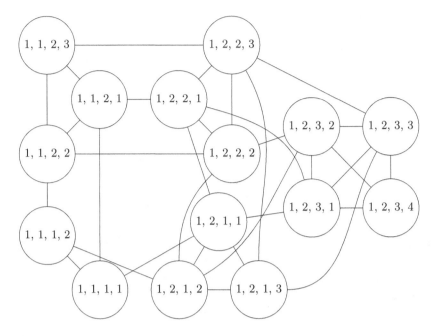

Fig. 6. A graph of partitions for 4 reads

with a 2, the sub-path starts at the child clique's node that ends in 2, visits all nodes ending with 3 or higher, and ends at the node ending with 1. Figures 5b and 7 show an example of extending a path in the graph of 3 reads to the graph of 4 reads. The graph does not need to be explicitly created, and only the nodes in the path are processed. Since extending the graph by one read will at least double the number of nodes, the total number of nodes processed is at most twice the number of nodes in the final graph, so ordering the partitions with the Gray code can be done in linear time in the number of partitions.

The algorithm has running time of $O(|S|C\frac{k^C}{k!})$, where C is maximum coverage, k is the number of strains and $|S|$ is the number of SNPs.

4   Experiments

The experiments were run on simulated data based on the *E. coli* genome. The simulated mutant genomes were created by taking the first ten thousand bases from the *E. coli* genome and creating four mutant strains from it. The mutations were created with a uniform 1 % probability of substitution per base. Only SNPs with substitutions were created. Reads were sampled from random locations with an average coverage of 20 for each strain, for a total average coverage of 80 before coverage reduction. All reads were created error-free.

Two different methods were used to build the haplotyping matrix from the reads. In the first method, the *simulated* method, the SNPs were directly read

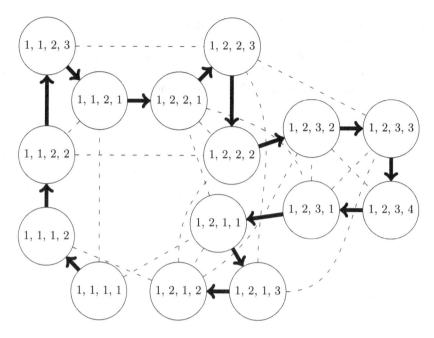

Fig. 7. A hamiltonian path in the partition graph over 4 reads

from the mutated genomes and the haplotyping matrix was built with full knowledge of the genomes. The simulated method represents a very optimistic upper bound for the algorithm's accuracy. In the second method, the *DiscoSNP* method, SNPs were detected by DiscoSNP [22] and processed to form the haplotyping matrix. The DiscoSNP method represents a more realistic impression of the algorithm's accuracy.

Two experimental settings were used. First, the read length was varied from 100 bases to 2000 bases, and the algorithm's accuracy was measured with each length. In the second setting, errors were introduced into the haplotyping matrix and the algorithm's accuracy measured for several read lengths. The errors were introduced directly into the haplotyping matrix immediately after creating it, before any coverage reduction. The reads passed to DiscoSNP were still error-free. This was done to make sure that the experiments only measure errors made by the implementation instead of the external tools.

We used *switch distance* to measure the accuracy of predictions. Originally switch distance is defined for two strains [13]. Informally, switch distance builds consensus genomes from its reads, and then aligns the consensus genomes to the actual genomes. Switch distance is then the number of times a consensus genome switches from one actual genome to another, allowing up to a certain number of alignment errors where a consensus genome's nucleotide differs from the actual genome's nucleotide.

To calculate the switch distance for multiple strains, we developed a simple dynamic programming algorithm, whose details can be found in [18]. Informally, this extended measure calculates the fraction of positions where genomes are switched. In the graphs, we report *switch accuracy*, which is the inverse of the switch distance, or the fraction of SNPs where genomes are not switched.

4.1   Effect of Varying Read Length

In the read length experiment, the implementation's accuracy was measured with varying read lengths. Read lengths were varied from 100 bases to 2000 bases. Figure 8 shows the switch accuracy.

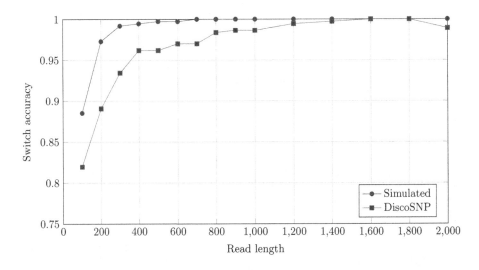

Fig. 8. Switch accuracy as a function of read length

The results show that accuracy depends greatly on read length. Even in the simulated method, reads shorter than about 300 bases have poor accuracy, and 100 bases long reads have an accuracy about as good as random guessing. On the other hand, long reads have a very high accuracy. The simulated method was completely accurate at 700 bases long, and the DiscoSNP method at 1600 bases, except for a strange dip at 2000 bases long reads. The experiment shows that the implementation cannot work for short reads even in the best case, but works well for long reads.

4.2   Effect of Varying Error Rate

In the error rate experiment, the implementation's accuracy was measured with varying error rate. The error rate was varied from 0 % to 15 % chance of a substitution error per base. The errors were introduced into the haplotyping matrix immediately after building the matrix, before any coverage reduction. Read lengths were 1000, 1600 and 2000 bases long for the DiscoSNP method, and 500, 700 and 1000 bases long for the simulated method. Only uniformly distributed substitution errors were considered. The switch accuracy stayed over 0.98 accuracy level in all these settings, and the increase in read length improved the accuracy level to 1 still at error level 8 % in simulated data.

5   Conclusion

In this paper we gave a proof of concept implementation of an exact haplotyping algorithm for multiple strains. The performance on simulated data is promising.

Acknowledgements. This work was supported in part by the Academy of Finland (grants 267591 to L.S. and 284598 (CoECGR)).

References

1. Aguiar, D., Istrail, S.: Haplotype assembly in polyploid genomes and identical by descent shared tracts. Bioinformatics **29**(13), i352–i360 (2013). http://bioinformatics.oxfordjournals.org/content/29/13/i352.abstract
2. Astrovskaya, I., et al.: Inferring viral quasispecies spectra from 454 pyrosequencing reads. BMC Bioinform. **12**(Suppl. 6), S1 (2011)
3. Bayzid, S., et al.: HMEC: a heuristic algorithm for individual haplotyping with minimum error correction. ISRN Bioinform. **2013**, 10 (2013)
4. Berger, E., et al.: Haptree: a novel bayesian framework for single individual polyplotyping using NGS data. PLOS Comput. Biol. **10**(3), e1003502 (2014)
5. Chen, Z., Deng, F., Wang, L.: Exact algorithms for haplotype assembly from whole-genome sequence data. Bioinformatics **29**(16), 1938–1945 (2013)
6. Cilibrasi, R., van Iersel, L., Kelk, S., Tromp, J.: On the complexity of several haplotyping problems. In: Casadio, R., Myers, G. (eds.) WABI 2005. LNCS (LNBI), vol. 3692, pp. 128–139. Springer, Heidelberg (2005). http://dx.doi.org/10.1007/11557067_11
7. Correll, D.S.: The Potato and Its Wild Relatives. Texas Research Foundation, Renner (1962)
8. Das, S., Vikalo, H.: SDhaP: haplotype assembly for diploids and polyploids via semi-definite programming. BMC Genomics **16**(1), 260 (2015). http://dx.org/10.1186/s12864-015-1408-5
9. Deng, F., Cui, W., Wang, L.: A highly accurate heuristic algorithm for the haplotype assembly problem. BMC Genomics **14**(Suppl. 2), S2 (2013)
10. He, D., et al.: Optimal algorithms for haplotype assembly from whole-genome sequence data. Bioinformatics **26**(12), i183–i190 (2010). http://bioinformatics.oxfordjournals.org/content/26/12/i183.abstract
11. Junttila, E.: Patterns in permuted binary matrices. Ph.D. thesis, University of Helsinki (2011)
12. Kuleshov, V.: Probabilistic single-individual haplotyping. Bioinformatics **30**(17), i379–i385 (2014). http://bioinformatics.oxfordjournals.org/content/30/17/i379.abstract

13. Lin, S., et al.: Haplotype inference in random population samples. Am. J. Hum. Genet. **71**(5), 1129–1137 (2002)

14. Lippert, R., et al.: Algorithmic strategies for the single nucleotide polymorphism haplotype assembly problem. Briefings in Bioinform. **3**(1), 23–31 (2002). http://bib.oxfordjournals.org/content/3/1/23.abstract

15. Mäkinen, V., et al.: Interval scheduling maximizing minimum coverage. CoRR abs/1508.07820 (2015). http://arxiv.org/abs/1508.07820

16. Neigenfind, J., et al.: Haplotype inference from unphased SNP data in heterozygous polyploids based on SAT. BMC Genomics **9**, 356 (2008)

17. Patterson, M., et al.: Whatshap: weighted haplotype assembly for future-generation sequencing reads. J. Comput. Biol. **22**(6), 498–509 (2015)

18. Rautiainen, M.: Identification of variant compositions in related strains without reference. Master's thesis, University of Helsinki (2016)

19. Stephens, J.C., et al.: Haplotype variation and linkage disequilibrium in 313 human genes. Science **293**(5529), 489–493 (2001). http://www.sciencemag.org/content/293/5529/489.abstract

20. Su, S.Y., et al.: Inference of haplotypic phase and missing genotypes in polyploid organisms and variably copy number genomic regions. BMC Bioinform. **9**, 513 (2008)

21. Tewhey, R., et al.: The importance of phase information for human genetics. Nat. Rev. Genet. **12**, 215–223 (2011)

22. Uricaru, R., et al.: Reference-free detection of isolated SNPs. Nucleic Acids Res. **43**(2), e11 (2014)

New Error Tolerant Method for Search of Long Repeats in DNA Sequences

Sergey P. Tsarev[1] and Michael G. Sadovsky[2(✉)]

[1] Siberian Federal University, Kirenskogo, 26, 660074 Krasnoyarsk, Russia
`sptsarev@mail.ru`
[2] Institute of Computational Modelling of SB RAS,
Akademgorodok, 660036 Krasnoyarsk, Russia
`msad@icm.krasn.ru`
`http://ikit.sfu-kras.ru`, `http://icm.krasn.ru`

Abstract. A new method to identify all sufficiently long repeating nucleotide substrings in one or several DNA sequences is proposed. The method based on a specific gauge applied to DNA sequences that guarantees identification of the repeating substrings. The method allows the matching substrings to contain a given level of errors. The gauge is based on the development of a heavily sparse dictionary of repeats, thus drastically accelerating the search procedure. Some biological applications illustrate the method.

Keywords: Genome · Fast search · Vernier pattern

1 Introduction

The classic problem of the search for the longest common string in two symbol sequences has a long story [1,2,6,8,9,11]. In spite of a number of deep and valuable results [3–5,7,10] obtained in the algorithm implementation for the problem, it is computationally challenging and an extremely active research field.

In brief, the problem we address here is the following. Let one (or more) sequences $\mathfrak{T}_1$, $\mathfrak{T}_2$, ..., $\mathfrak{T}_k$ from some finite alphabet are given; further we shall concentrate on the four-letter alphabet $\aleph = \{A, C, G, T\}$ only, since we illustrate the results with genetic data. So, the problem is to find all sufficiently long substrings $\{\mathfrak{s}_i\}$ that occur at least twice in $\mathfrak{T}_i$.

The problem could be understood in two different ways: the former is a search for the **exactly** matching substrings, and the latter is a search for two substrings bearing some tolerable mismatches; obviously, the first problem is a special case of the second one. Section 2 describes a primitive search algorithm meeting the exact match constraint; Sect. 3 presents our main idea of much faster search method which additionally allows an expansion for approximate matching case. In Sect. 4 we give a brief survey of experimental verifications of our method. In Sect. 5 an important combinatorial problem related to the proposed method is discussed.

© Springer International Publishing Switzerland 2016
M. Botón-Fernández et al. (Eds.): AlCoB 2016, LNBI 9702, pp. 171–182, 2016.
DOI: 10.1007/978-3-319-38827-4_14

The new method has the following advantages:

- it is much more economic in comparison to exhaustive search for all repeating substrings of an arbitrary length. We search for all repeats longer than a given integer N (provided by researcher). Greater N accelerates our algorithm;
- it finds simultaneously all repeats in a given DNA sequence (or in any other string of symbols in any finite alphabet) or common substrings in two or more symbol sequences;
- it permits an error tolerance: a portion of mismatches in compared substrings is allowed. Although the current implementation does not guarantee all repeats (or common substrings) of the length N or greater with given tolerance level of mismatches identification, test runs have shown that the probability of missing of inexact repeats with the given tolerance is small. Still the current implementation guarantees all exact repeats to be found, cf. the discussion in Sect. 3.4.

The problem actuality results in a tremendous growth of the papers devoted to it. There is a number of various algorithms to resolve the problem, and the number of software implementations falls beyond imagination; here we have no chance even to enlist all of them, due to space limitation. Nonetheless, some related results and reference details could be found in [3–5, 7, 10].

The authors thank Prof. S.V. Znamenskij for useful discussions; the idea of Vernier gauge for acceleration of search was also independently found by him.

2   Long Repeats by Brute Force

Here we sketch a well-known primitive but exhaustive algorithm to search repeating substrings in symbol strings. This algorithm is still of practical use for analysis of DNA sequences as long as 10^7 or so, and can be used later to check more advanced algorithms described below.

Theoretically the problem of searching for a repeating substring may be reduced to a construction of a frequency dictionary for the given symbol sequence $\mathfrak{T}$; the former is a list of all the substrings (of the given length m, also called "thickness" of dictionary) occurred within the sequence $\mathfrak{T}$ so that each entry in the dictionary is associated with the frequency of the relevant string in $\mathfrak{T}$. A dictionary W_m (of the thickness m) could be defined in a variety of ways; cf. for example [12] (where it is called *finite dictionary*).

The simplest way to develop W_m is as follows. Let us fix a window of the length m that identifies a substring in a sequence $\mathfrak{T}$, and a step t for the window shift alongside the sequence. Thus, the frequency dictionary $W_{m,t}$ is the set of all the strings of the length m identified by the window of that length moving alongside the sequence with the step t. Each element of the dictionary is assigned with its frequency (the number of copies of this element met in the dictionary building process) and (for our purposes) the list of all positions where the given element of the dictionary has been met.

Having $W_{m,1}$, one easily can find all the repeats of the length $N \geq m$ in $\mathfrak{T}$, selecting all elements s of $W_{m,1}$ that are met more than in one copy and (using the list of the positions of s) clustering all other repeating elements of $W_{m,1}$ with consecutive position tags.

The question whether two (or several) sequences $\mathfrak{T}_1$ and $\mathfrak{T}_2$ have a common substring s of the length N could be addressed through a comparison of the frequency dictionaries of those sequences. In our test runs we have built up dictionaries of thickness $m \leq 10000$ for a single DNA sequence of length $44 \cdot 10^6$ base pairs (*Bos taurus* chromosome 25). The dictionaries of the given thickness were built in three stages:

1. first, we identified substrings of the given length m with step $t = 1$ and develop an intermediate "predictionary" text file `F.predic` where each substring occupies a separate line and is accompanied by the position tag;
2. second, `F.predic` is sorted lexicographically using the standard system command `sort`;
3. third, the identical substrings in the sorted file are eliminated so that the resulting line bears the substring, accumulated number of its copies and the list of position tags gathered from the eliminated substrings in the sorted `F.predic`.

Step 2 (lexicographical sorting) is the most time consuming. It could be executed in reasonable time on a mainframe with 30 Gb of RAM under OS Linux (http://cluster.sfu-kras.ru/page/supercomputer/). The following table shows the run time of this step for several m values (Table 1).

Table 1. Runtime of the tests for brute-force dictionary development; t_s is the sorting time, m is a substring length.

m	`F.predic` size	t_s	m	`F.predic` size	t_s
200	8.3 Gbytes	10 min	1000	40 Gbytes	17 min
500	20 Gbytes	11 min	10000	400 Gbytes	1 h 12 min

Hence, the steps 1–3 yield the following results, in terms of the frequency dictionary structure. For $m = 200$ the observed number N_k of (different) strings (of the length m) met in k copies is following:

k	2	3	4	5	6	7	8	9	≥ 10
N_k	312338	3600	756	203	72	2	0	0	80

Similar figures for $m = 500$ are

k	2	3	4	5	6	7	8	9	≥ 10
N_k	252338	126	14	33	0	0	0	0	33

For $m = 1000$ $N_2 = 193227$ strings were met in two copies, and none has been found to be met in three or greater number of copies. Similarly, for $m = 10000$ these figures were $N_2 = 18865$ and $N_{>2} = 0$, respectively. In fact all repeats of substrings of length 10000 in this DNA sequence were clustered on step 3 into 3 exactly matching substrings of lengths (approximately) 11000, 15000 and 21700.

3   Vernier Gauge Algorithm

Here we introduce a new much faster method to search for *a repeating substring of the length $\geq N$ in a symbol sequence (alternatively, a common substring in several symbol sequences)*. The key idea is to change the analysis of a complete frequency dictionary $W_{m,1}$ (where each symbol in the sequence $\mathfrak{T}$ gives a start to a string of the length m) to the analysis of sparse frequency dictionary W_{m,t_s} with *variable step* t_s having significantly less number of entries. An idea standing behind the proposed method is strongly connected to a well-known Vernier scale [13] used to measure length with enhanced precision in comparison to the standard scale.

3.1   Simple Example

In this subsection we develop the idea of the *simplest* Vernier gauge to search a common substring of length N or more in two symbol sequences. Simply speaking, we should cover the first sequence with *tags* of some small length m with some step k; the second sequence must be covered with *tags* of the same length, but here the step between (the beginning letters of) two neighboring tags is equal to $k - 1$, and not k. If a tag is found in both sequences, it must be examined for *expansion* (see below). Let us give a closer look at this process.

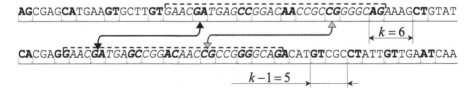

Fig. 1. Illustration of the Vernier approach to find out a sufficiently long common substring in two sequences.

Suppose two sequences $\mathfrak{T}_1$ and $\mathfrak{T}_2$ have a common substring $\mathfrak{s}$ of the length N; yet, we have no idea about the locations of that common substring in the sequences. Let us build two frequency dictionaries: $W^1_{m,k}$ and $W^2_{m,k-1}$ with $k \leqslant \sqrt{N}$ and some thickness m for $\mathfrak{T}_1$ and $\mathfrak{T}_2$ respectively. The second dictionary bears the strings of the same length m. Both dictionaries must start from the very beginning of each sequence.

It should be stressed that parameters k and m are almost independent (cf. Theorem 1 for a precise statement). The choice of these latter is determined by the (expected) length N of a common substring, while m is to be chosen almost arbitrary: it would be nice, if the frequency dictionary of the tags is almost degenerated (i.e. the greatest majority of the tags should exist in few copies). Figure 1 illustrates this idea for $k = 6$, $m = 2$. The common substring of length $N = 31$ is indicated in slanted font, in the figure. The Vernier gauge (see Sect. 3.3 for details) identifies the common short sub-substring GA in both sequences; they are indicated with the curve arrow in black. Incidentally we can see one more occurrence of a common sub-substring CG of length 2 in out target common substrings; this is not a typical case, but we shall keep in mind such a possibility, as well.

A closer inspection gives three more common sub-substring CA, CT and GT of length 2 in the dictionaries $W_{2,6}^1$ and $W_{2,5}^2$. They do not *expand* to a common target substring of length $N = 5 \times 6 + 2 - 1 = 31$ in $\mathfrak{T}_1$ and $\mathfrak{T}_2$ (more on *expansion* of the common entries in the dictionaries $W_{m,k}^1$ and $W_{m,k-1}^2$ see below and in Subsect. 3.4). This abundance of common entries in $W_{2,6}^1$ and $W_{2,5}^2$ results from the small capacity of the nucleotide alphabet ACGT and small length $m = 2$ chosen for this simple example.

The idea of searching the common substring using relatively very short tags and rarefied dictionaries (we call it the *double Vernier gauge* on $\mathfrak{T}_i$) is based on the following simple theorem.

Theorem 1. *If there is a common substring of the length N or more in $\mathfrak{T}_1$ and $\mathfrak{T}_2$ then a common entry (sub-substring of length m) can be found in dictionaries $W_{m,k}^1$ and $W_{m,k-1}^2$ developed for sequences $\mathfrak{T}_1$ and $\mathfrak{T}_2$, respectively, provided that $N \geq k(k-1) + m - 1$.*

Proof. Let s_1 and s_2 be the common (exactly matching) substrings of the length $N \geq k(k-1) + m - 1$ starting at positions u and v in $\mathfrak{T}_i$ respectively. First, we cut off (virtually, for simplicity of the proof) their last $m - 1$ symbols and look only at the starting positions of the tags of the dictionaries $W_{m,k}^i$ in s_i. Let $0 \leq \alpha < k$ and $0 \leq \beta < k - 1$ be the starting positions of the dictionary entries in s_1 and s_2 *relative to their starting symbols* respectively. The other starting positions of the dictionary tags inside s_1 will be $\alpha + x \cdot k$ with $x \in \{0, 1, \ldots, k-2\}$, for s_1 and $\beta + y \cdot k$ with $y \in \{0, 1, \ldots, k - 1\}$, for s_2 (w.r.t. the starting symbols of s_i). We should find such integers x, y that $\alpha + x \cdot k = \beta + y \cdot (k - 1)$, i.e. $x \cdot k - y \cdot (k - 1) = \beta - \alpha$ with the constraints on x, y given above. In fact, if $\gamma = \beta - \alpha \geq 0$, $x = y = \gamma$ is the solution. For $\gamma = \beta - \alpha < 0$, $x = \gamma + k - 1$, $y = \gamma + k$ should be taken. $\square$

In order to find duplicate tags in the dictionaries one may apply different standard techniques; in our current simplest implementation standard lexicographic sorting and merging of $W_{m,k}^1$ and $W_{m,k-1}^2$ are used. As soon as all common tags in the dictionaries $W_{m,k}^1$ and $W_{m,k-1}^2$ are found, the next steps of our algorithm are:

- *expand* the found common tags using their positions in $\mathfrak{T}_i$. Namely consecutively compare the symbols on the right of the tags in $\mathfrak{T}_1$ and $\mathfrak{T}_2$ as far as they match, stopping when we meet non-matching symbols. Then consecutively compare the symbols on the left of the tags in the same way as far as they match.

– If the length of the *expanded tag* is at least N, add it to the list of successful expansions for further output after all identical tag pairs in $W^1_{m,k}$ and $W^2_{m,k-1}$ are *expanded*.

In the simplest version of our algorithm discussed in this subsection we are searching only *exactly matching* substrings. Also we do not take into account the possibility to meet some other symbols than the standard nucleotides A, C, G, T. If one expects that some other symbols (like N, W etc.) may occur in the analyzed DNA sequences then one of the expansion strategies discussed in Sect. 3.4 should be applied. In the next subsection we consider the problem of choice of the parameter m.

3.2   Tag Length Choice to Enforce Vernier Gauge Algorithm

The sub-substrings of length m chosen to build up the dictionaries $W^1_{m,k}$ and $W^2_{m,k-1}$ in the previous subsection are called **tags**. Its length m is very important parameter affecting speed and overall efficiency of our algorithm. A smart choice of this parameter may dramatically reduce the processing time and, what is even more important, is crucial in the process of subsequent *expansion* of the tags common for both dictionaries $W^1_{m,k}$ and $W^2_{m,k-1}$ into the full common strings of the length $\geq N$ *with the given proportion of errors* in the last stage of the algorithm execution (Sect. 3.3).

In fact, the capacity of the nucleotide alphabet dictates the choice of sufficiently large m to minimize the number of sporadic coincidences of tags in the dictionaries. The experiments in Sect. 4 show that $m = 30$ is good enough, when the steps k, $k - 1$ are greater than 30 (so the target length of common substrings sought by the algorithm is at least 1000). It should be stressed that overlapping of the tags in $\mathfrak{T}_i$ (if $m > k - 1$) is not a problem, in our approach; overlapping itself does not affect the algorithm.

3.3   General Description of the Problem

The double Vernier gauge described in Sect. 3.1 stands behind the more general search pattern presented below in Sects. 5.1 and 5.2. Here we will discuss only the simplest modifications of the double Vernier gauge necessary to find repeats in one or several DNA sequences.

The general problem solved by our algorithms is:
given parameters N (an integer) and ϵ (a positive real number), find all substrings of the length at least N in one or several sequences $\mathfrak{T}_i$ that occur repeatedly (exact matching requires $\epsilon = 0$) or couples of substrings in $\mathfrak{T}_i$ that differ at most at q places, $q = \epsilon \cdot \text{length}(s)$.

3.4   How the Method Works

Step 1. Given the target length N, choose proper k and m such that $N \geq k(k-1) + m - 1$.

Step 2. If we have two DNA sequences to analyze, develop the dictionaries $W^1_{m,k}$ and $W^2_{m,k-1}$. Otherwise (for one or more than two DNA sequences) develop for each DNA sequence a dictionary with variable step: take tags of length m starting at positions $1, k, k+1, 2(k-1)+1, 2k+1, \ldots$ — that is at the union of the subsets $\{p \cdot (k-1) + 1, p = 0, 1, 2, \ldots\}$ and $\{q \cdot k + 1, q = 0, 1, 2, \ldots\}$. Add positions of the selected tags into the dictionaries.

Step 3. Check whether there may be common entry tags in the dictionaries. If we want to find repeats in one DNA sequence (or all possible repeats in several DNA sequences $\mathfrak{T}_i$, possibly the same ones), find repeated tags in one dictionary or in the dictionary merged from all dictionaries built for all $\mathfrak{T}_i$. Strategies for finding common tag entries are discussed below.

Step 4. Expand the found repeated tags (using their positions stored in the dictionaries) as described in Sect. 3.1 if the exact matching is required. If a positive tolerance $\epsilon > 0$ is given, use one of the expansion strategies discussed below.

Step 5. List all expanded tags with their positions in $\mathfrak{T}_i$. If some of the expanded substrings are shorter than N one may keep or reject them (there is no guarantee that all matching substrings of length $< N$ will be found!)

More Technical Details. *Finding common tags on Step 3.* In our current implementation we use the standard lexicographic sorting of the tags (using the standard system command sort) and merge the sorted dictionaries. The positions of the tags stored in the dictionaries are added to the tag on the same line after a space so identical tags will be on two or more consecutive lines after sorting. Then using any text processing utility (for example the standard gawk) we find such consecutive lines with identical tags and build a list of repeated tags with their positions in the respective $\mathfrak{T}_i$. This is rather fast for the examples described in Sect. 4, but if the error tolerance ϵ is positive (so inexact matches are to be allowed) lexicographic sorting does not guarantee that we will find all tags that match inexactly with the given tolerance level ϵ.

Expansion strategies on Step 4. If $\epsilon = 0$ simple expansion described in Sect. 3.1 should be applied, that is compare consecutively the symbols on the right and on the left of the identical tags in $\mathfrak{T}_i$ as far as they match, stopping when we meet non-matching symbols. If $\epsilon > 0$ then during this process continue expansion even if the compared symbols near the tags do not match; if such a non-match is found add 1 to the counter miss_count of mismatches and stop the process of expansion if miss_count/length(s) $> \epsilon$ (here s is the string obtained in the process of expansion).

Treatment of symbols N, W *etc.* In the currently available DNA databases one encounters results with non-exact recognition of nucleotides, they are marked by letters outside of the standard nucleotide alphabet $\aleph = \{A, C, G, T\}$. Several strategies may be applied depending on the problem solved by the researcher:

- consider symbols N, W etc. as *errors* adding 1 to miss_count;
- consider them as *possible matches* (not recognized by the DNA sequencer) and keep expansion without adding 1 to miss_count.
- cut the DNA sequences into smaller pieces which do not contain such extra symbols and run our algorithm on the obtained pieces.

4   Preliminary Experimental Results

We checked the developed algorithm over the following genetic data (all sequences were retrieved from EMBL–bank):

(1) *Human chromosome 14* (since it contains A, C, G, T symbols only);
(2) 4 sets of drosophila genomes:
    - *Drosophila melanogaster,*
    - *Drosophila simulans,*
    - *Drosophila simulans* strain white501,
    - *Drosophila yakuba* strain Tai18E2;
(3) *Bos taurus* complete genome.

4.1   Human Chromosome 14

When the algorithm of Sect. 3.4 was run with the parameters $m = 50$, $k = 31$ (so we find all repeats of length at least $N = k(k - 1) + m - 1 = 979$, in total 19946 repeated tags were found on Step 3, among them 12154 tags occur twice, 3670 tags occur thrice, ..., 12154 tags occur 10 times or more, the maximal frequency was 25.

After expansion on Step 4 with $\epsilon = 0$ (only exact matching was allowed) two identical substrings of length 1019 were found, as well as hundreds of repeats of smaller lengths.

After expansion on Step 4 with $\epsilon = 0.02$ a pair of approximately matching substrings of length 11000 (we give the approximate length since due to our expansion method there are a few dozens of mismatches at the both ends of them) was found, as well as few hundreds of repeats (approximate matches) of lengths 1000 and more. The pair of length >11000 is in fact a long almost periodic subsequence with period 102: the first string of the found pair starts from position 85597640 and the second one is shifted to the end of the chromosome by 102 positions. If one compares these approximately matching substrings then one sees a few *exactly matching sub-substrings* of the following lengths (given in the order of appearance, only exact matches of length more than 10 are given): 38, 101, 305, 203, 468, 101, 652, 101, 298, 55, 38, 62, 242, 94, 196, 101, 101, 203, 108, 305, 203, 305, 101, 101, 94, 101, 196, 94, 101, 62, 38, 62, 344, 62, 38, 62, 140, 62, 38, 62, 242, 101, 196, 94, 101, 24, 123, 52, 24, 21, 77, 287, 94, 196, 203, 62, 38, 62, 147, 101, 157, 203, 305, 101, 45, 101, 203, 94, 196, 713, 101, 62, 101, 38, 62, 38, 164. Many of them are multiples of the period 102 minus 1. Typically only 1 mismatching nucleotide occurs between the exactly matching subregions.

4.2   4 Sets of Drosophila Genomes

The total size of this genome collection of 24 chromosomes of all 4 species is $478 \cdot 10^6$ BP. It contains several millions of unrecognized nucleotides (marked with n symbol, mostly met in the last 3 species). When the algorithm of Sect. 3.4 was run on the complete set of all chromosomes with the parameters $m = 50$, $k = 63$ (so we find all repeats of length at least $N = 3995$), in total 180980 repeated tags of length m were found on Step 3, the maximal frequency was 34.

We have tried a different choice of $m = 50$, $k = 200$ (so we find all repeats of length at least $N = 40000$), which produced 20442 repeated tags (maximal frequency 12). On Step 4 we treated the n symbols as errors. When treated separately, the 4 species exhibit considerable differences in repeat lengths and the overall number of long repeats ($\epsilon = 1/50$):

- *Drosophila melanogaster* genome has 9 repeats of length 10000 and more, the longest being 30893.
- *Drosophila yakuba* strain Tai18E2, *Drosophila simulans* and *Drosophila simulans* strain white501 genomes do not have such long exact repeats (but they have dozens of exact repeats of length 1000 and more with maximal length 3024). On the other hand these repeats expand to approximate repeats of length up to 6000 if processed with $\epsilon = 1/50$.

4.3   *Bos Taurus* Complete Genome

The overall size of 29 processed DNA sequences of the complete genome was more than $2.4 \cdot 10^9$ symbols. Since the files contain large unrecognized nucleotide substrings a number of different strategies described in Sect. 3.4 were tried. If the n symbols (no other unrecognized symbols were encountered) were considered as non-errors, huge repeats of length up to 300000 were found; they consists of n symbols practically completely. When the files were cut into pieces not containing n symbols (this resulted in 9718 files of size greater than 100 Kbytes and 11000 smaller files) and processed with the parameters $m = 50$, $k = 600$ (thus $N > 360000$) a number of exact repeats of length up to 89453 were found.

5   Discussion

The experiments described in the previous Section show that our method is sufficiently fast and yields the results interesting both for exact and approximate sequence analysis. Still, a number of questions arises concerning a feasibility of the method for various biologically meaningful issues; a search for degenerated motifs is among them. First, here we present a theoretical result rather than a ready-to-use software package. Our implementation aims just to check a feasibility of the method itself. Evidently, there is no obstacles to combine, in some way, Vernier sparse search and other well-established techniques (suffix trees, etc.).

Second, the current implementation guarantees revealing of all *exact* matches; if tolerance level $\epsilon > 0$, some minor changes must be implemented to avoid a failure of the method caused by the coincidence of a tag with admissible mis-matches in degenerated motif. Indeed, simple lexicographically arranged sorting of tags (sparse dictionary entries) must be changed for the search of tags that are close with respect to admissible mismatch patterns (Levenstein distance, edit distance, etc.).

Finally, a correct comparison of the speed of execution of software realizing Vernier method and the combinations of that latter with some other approaches should be done explicitly; a lack of space here blocks us to do that. An extended version of this paper could be found on www.arXiv.org/archive/q-bio later.

In addition, some interesting mathematical and algorithmic issues are to be urged to optimize the process. The following subsections address the issues.

5.1   Circular and Linear Vernier Patterns

Actually, the procedure described in Sect. 3 implies the following property of the standard double Vernier gauge:

suppose some positions i_1, i_2, ..., i_k in N-element set $\mathfrak{N} = \{1, 2, \ldots, N\}$ are marked. Then, if these marks are periodically repeated in a larger set $\mathfrak{M} = \{1, 2, \ldots, M\}$, $M \gg N$, then for any $s < M - N$ one finds at least two marked positions with the exact distance s between them.

The property guarantees that for any two identical substrings $\mathfrak{s}_1$, $\mathfrak{s}_2$ of length $N + m - 1$ located in a longer symbol sequence $\mathfrak{T}$ of length $M + m - 1$ (the starting positions of $\mathfrak{s}_1$ and $\mathfrak{s}_2$ differ in s symbols), couple of the marked positions exists in $\mathfrak{T}$ in the same position in $\mathfrak{s}_1$, $\mathfrak{s}_2$ with respect to their starting symbols, so the tags (sub-substring of length m) starting at the selected positions inside $\mathfrak{T}$ coincide.

A better geometric insight into this **Vernier pattern** of positions i_1, i_2, ..., i_k is given by the following construction:

taking a circumference of length N and starting from some point O (corresponding to the element 1 in the set $\mathfrak{N}$), mark clockwise the points at the distances $i_1 - 1$, $i_2 - 1, \ldots, i_k - 1$ from O on the circumfer-ence. Then for any integer length $s \leq N/2$ one finds at least two marks spanning the (shortest) arc of length s.

The circular picture corresponds to periodic repetition of the marks in the larger M-element set $\mathfrak{M}$. If for the given integer N one finds set of positions $\mathfrak{V} = \{i_1, i_2, \ldots, i_k\}$ (with integer elements $0 < i_p \leq N$) satisfying the property formulated above, then such $\mathfrak{V}$ is called an (N, k)-**circular Vernier pattern.**

Another concept of Vernier pattern may be introduced. It is referred to sim-plified version of the problem of DNA sequence assembly. Namely, if for a subset $\mathfrak{V} = \{i_1, i_2, \ldots, i_k\} \subset \mathfrak{N}$ and for any integer length $s \leq N$ one finds at least two elements in $\mathfrak{V}$ with the distance s between them, then such $\mathfrak{V}$ is called (N, k)-**linear Vernier pattern.**

5.2   Minimalistic and Minimal Vernier Patterns

Obviously, the smaller k for given N is taken, the more economic dictionary could be developed using the tags with the starting positions at the elements of an (N, k)-circular Vernier pattern $\mathfrak{V}$ periodically repeated in a large DNA sequence $\mathfrak{T}$. Since the number of different distances between k points can not be greater than $k(k-1)/2$, we have the following lower bound for k: $k(k-1)/2 \geq N/2$, thus for big N, $k \sim \sqrt{N}$. So, for the double Vernier gauge described in Subsect. 3.1 for a search of repeats in *two* DNA sequences, we have in fact a minimal possible choice of marks (beginning positions of the tags). For all other cases described as Step 2 of our algorithm in Subsect. 3.4, we have approximately twice more marked positions in each of the $\mathfrak{T}_i$. For linear Vernier patterns the situations is slightly different: $k(k-1)/2 \geq N_1 = N - m + 1$.

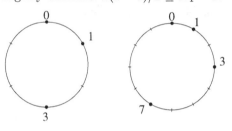

Fig. 2. An example of minimalistic circular Vernier patterns, for $k = 3$ (left) and $k = 4$ (right).

So the following mathematical problem is a good combinatorial challenge.

Problem $\mathfrak{V}$. *For any given integer N find circular and linear Vernier patterns with minimal possible k.*

Such Vernier patterns are called *minimal Vernier patterns*. For small N one can even find *minimalistic* Vernier patterns, i. e. the patterns with $k(k-1)/2 = \lfloor N/2 \rfloor$ (resp. $k(k-1)/2 = N_1$ for linear patterns). Here on Fig. 2 we give some examples of minimalistic circular Vernier patterns.

Acknowledgments. This study was supported by a research grant No. 14.Y26.31. 0004 from the Government of the Russian Federation (M.G. Sadovsky) and the grant from Russian Ministry of Education and Science to Siberian Federal University, contract *No.* 1.1462.2014/K (S.P. Tsarev).

References

1. Altschul, S.F., Gish, W., Miller, W., Meyers, E.W., Lipman, D.J.: Basic local alignment search tool. J. Mol. Biol. **215**(3), 403–410 (1990)
2. Bergroth, L., Hakonen, H., Raita, T.: A survey of longest common subsequence algorithms. In: SPIRE 2000, pp. 39–48. IEEE Computer Society (2000)
3. Chen, G.L., Chang, Y.J., Hsueh, C.H.: PRAP: an ab initio software package for automated genome-wide analysis of DNA repeats for prokaryotes. Bioinformatics. **29**(21), 2683–2689 (2013)
4. Erciyes, K.: Distributed and Sequential Algorithms for Bioinformatics. Computational Biology, vol. 23, p. 367. Springer, Cham (2015)
5. Girgis, H.Z.: Red: an intelligent, rapid, accurate tool for detecting repeats de-novo on the genomic scale. BMC Bioinform. **16**, 227 (2015)

6. Hirschberg, D.S.: A linear space algorithm for computing maximal common subsequences. Commun. ACM **18**(6), 341–343 (1975)
7. Lian, S., Chen, X., Wang, P., Zhang, X., Dai, X.: A complete and accurate Ab initio repeat finding algorithm. Interdiscip. Sci. **8**(1), 75–83 (2016)
8. Maier, D.: The complexity of some problems on subsequences and supersequences. J. ACM **25**(2), 322–336 (1978). ACM Press
9. Masek, W.J., Paterson, M.S.: A faster algorithm computing string edit distances. J. Comput. Syst. Sci. **20**(1), 18–31 (1980)
10. Novák, P., Neumann, P., Pech, J., Steinhaisl, J., Macas, J.: RepeatExplorer: a galaxy-based web server for genome-wide characterization of eukaryotic repetitive elements from next-generation sequence reads. Bioinformatics **29**(6), 792–793 (2013)
11. Pearson, W., Lipman, D.: Improved tools for biological sequence comparison. PNAS USA **85**, 2444–2448 (1988)
12. Sadovsky, M.G.: Information capacity of nucleotide sequences and its applications. Bull. Math. Biology. **68**, 156 (2006)
13. https://en.wikipedia.org/wiki/Vernier_scale

Author Index

Printed in the United States
by Baker & Taylor Publisher Services